my revision notes

WJEC GCSE

SCIENCE DOUBLE AWARD

Adrian Schmit
Jeremy Pollard

HODDER
EDUCATION
AN HACHETTE UK COMPANY

Acknowledgements

Exam practice questions at the end of each chapter in the physics section are reproduced by permission of WJEC.

Photo credits

p. 203 (top) © Washington Imaging/Alamy Stock Photo, (bottom) © ACORN 1/Alamy Stock Photo; p. 211 © sciencephotos/Alamy Stock Photo; p. 212 (left) © Berenice Abbott/Science Photo Library; (right) © Andrew Lambert Photography/Science Photo Library

The Publishers would like to thank the following for permission to reproduce copyright material.

Hachette UK's policy is to use papers that are natural, renewable and recyclable products and made from wood grown in well-managed forests and other controlled sources. The logging and manufacturing processes are expected to conform to the environmental regulations of the country of origin.

Orders: please contact Hachette UK Distribution, Hely Hutchinson Centre, Milton Road, Didcot, Oxfordshire, OX11 7HH. Telephone: +44 (0)1235 827827. Email education@hachette.co.uk. Lines are open from 9 a.m. to 5 p.m., Monday to Friday. You can also order through our website: www.hoddereducation.co.uk

Cover photo © NATUREWORLD/Alamy Stock Photo

Typeset in Bembo Std Regular 11/13 by Integra Software Services Pvt. Ltd, Pondicherry, India

Printed and bound by CPI Group (UK) Ltd, Croydon, CR0 4YY

A catalogue record for this title is available from the British Library.

MIX
Paper | Supporting
responsible forestry
FSC™ C104740
FSC
www.fsc.org

Get the most from this book

Everyone has to decide his or her own revision strategy, but it is essential to review your work, learn it and test your understanding. These Revision Notes will help you to do that in a planned way, topic by topic. Use this book as the cornerstone of your revision and don't hesitate to write in it — personalise your notes and check your progress by ticking off each section as you revise.

Tick to track your progress

Use the revision planner on pages iv to ix to plan your revision, topic by topic. Tick each box when you have:
- revised and understood a topic
- tested yourself
- practised the exam questions and gone online to check your answers and complete the quick quizzes

You can also keep track of your revision by ticking off each topic heading in the book. You may find it helpful to add your own notes as you work through each topic.

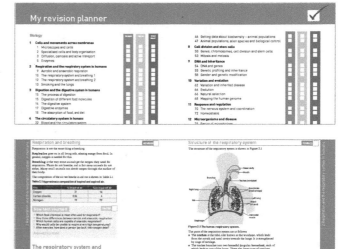

Features to help you succeed

Exam tips

Expert tips are given throughout the book to help you polish your exam technique in order to maximise your chances in the exam.

Now test yourself

These short, knowledge-based questions provide the first step in testing your learning. The answers are online (see below).

Definitions and key words

Clear, concise definitions of essential key terms are provided where they first appear.

Key words from the specification are highlighted in bold throughout the book.

ⓗ Where this symbol appears, the text to the right of it relates to higher tier material.

Exam practice

Practice exam questions are provided for each topic. Use them to consolidate your revision and practise your exam skills.

Summaries

The summaries provide a quick-check bullet list for each topic.

Online

Go online to check your answers to the Now test yourself and Exam questions and try out the extra quick quizzes at **www.hoddereducation.co.uk/ myrevisionnotesdownloads**

My revision planner

Biology

REVISED   TESTED   EXAM READY

REVISED   TESTED   EXAM READY

REVISED   TESTED   EXAM READY

33 Half-life

Now test yourself answers, Exam practice answers and quick quizzes at www.hoddereducation.co.uk/myrevisionnotesdownloads

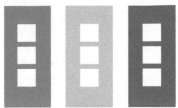

Countdown to my exams

6–8 weeks to go

- Start by looking at the specification — make sure you know exactly what material you need to revise and the style of the examination. Use the revision planner on pages iv to ix to familiarise yourself with the topics.
- Organise your notes, making sure you have covered everything on the specification. The revision planner will help you to group your notes into topics.
- Work out a realistic revision plan that will allow you time for relaxation. Set aside days and times for all the subjects that you need to study, and stick to your timetable.
- Set yourself sensible targets. Break your revision down into focused sessions of around 40 minutes, divided by breaks. These Revision Notes organise the basic facts into short, memorable sections to make revising easier.

REVISED ☐

2–6 weeks to go

- Read through the relevant sections of this book and refer to the exam tips, summaries, and key terms. Tick off the topics as you feel confident about them. Highlight those topics you find difficult and look at them again in detail.
- Test your understanding of each topic by working through the 'Now test yourself' questions in the book. Look up the answers online.
- Make a note of any problem areas as you revise, and ask your teacher to go over these in class.
- Look at past papers. They are one of the best ways to revise and practise your exam skills. Write or prepare planned answers to the exam practice questions provided in this book. Check your answers online and try out the extra quick quizzes at **www.hoddereducation.co.uk/ myrevisionnotesdownloads**
- Use the revision activities to try out different revision methods. For example, you can make notes using mind maps, spider diagrams or flash cards.
- Track your progress using the revision planner and give yourself a reward when you have achieved your target.

REVISED ☐

One week to go

- Try to fit in at least one more timed practice of an entire past paper and seek feedback from your teacher, comparing your work closely with the mark scheme.
- Check the revision planner to make sure you haven't missed out any topics. Brush up on any areas of difficulty by talking them over with a friend or getting help from your teacher.
- Attend any revision classes put on by your teacher. Remember, he or she is an expert at preparing people for examinations.

REVISED ☐

The day before the examination

- Flick through these Revision Notes for useful reminders, for example the exam tips, summaries and key terms.
- Check the time and place of your examination.
- Make sure you have everything you need — extra pens and pencils, tissues, a watch, bottled water.
- Allow some time to relax and have an early night to ensure you are fresh and alert for the examinations.

REVISED ☐

1 Cells and movements across membranes

Microscopes and cells

Plant and animal cells

All cells in both plants and animals have certain features in common:
- **Cytoplasm**, where most of the chemical reactions that make up life go on.
- A **cell membrane**, which controls what enters and leaves the cell.
- A **nucleus**, which contains DNA, the chemical that controls the cell's activities.
- **Mitochondria** (singular: mitochondrion), which are the structures that carry out aerobic respiration, supplying cells with energy.

Plant cells can be distinguished from animal cells, because they have some features that are not seen in animal cells. These are:
- A **cell wall**, made of cellulose.
- A large, **permanent central vacuole** filled with cell sap.
- **Chloroplasts**, which absorb light for photosynthesis – chloroplasts are not found in *all* plant cells, but they are never found in animal cells.

Figure 1.1 shows the structure of an animal and plant cell, and the differences between them.

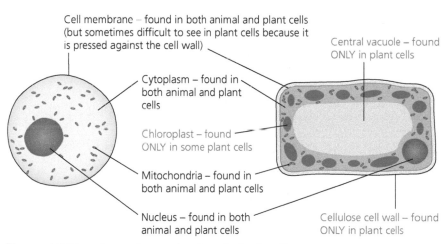

Cell membrane – found in both animal and plant cells (but sometimes difficult to see in plant cells because it is pressed against the cell wall)

Central vacuole – found ONLY in plant cells

Cytoplasm – found in both animal and plant cells

Chloroplast – found ONLY in some plant cells

Mitochondria – found in both animal and plant cells

Nucleus – found in both animal and plant cells

Cellulose cell wall – found ONLY in plant cells

Figure 1.1 Structure of an animal and plant cell.

Using a microscope

The parts of a microscope are shown in Figure 1.2.

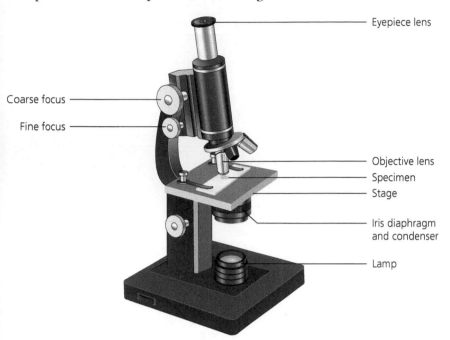

Figure 1.2 Parts of a microscope.

The functions of the parts of a light microscope are as follows:
- The **eyepiece lens**, which has a fixed magnification.
- The **objective lenses** are of different magnifying powers and are interchangeable, to adjust the magnification of the image that you see down the microscope.
- The **stage** is where the microscope slide is placed.
- Below the stage is a part that is usually made of up of two components – an **iris diaphragm**, which can be opened or closed to adjust the amount of light entering the objective lens, and (sometimes) a **condenser**, which concentrates the light into a beam directed precisely into the objective lens.
- At the base of the microscope is a **lamp**, or possibly a mirror, which is used to shine light through the condenser and iris diaphragm.
- The microscope is focused using two focus controls. The **coarse focus control** is used to get the image roughly into focus using the lowest-power objective, and then the **fine focus control** is used to fine tune the image and make it as clear as possible.
- Microscope **slides** hold thin specimens or sections, which may be stained using a variety of dyes so that structures can be seen more clearly.

Specialised cells and body organisation

Newly formed cells undergo a process of **differentiation** into **specialised cells**, which have features which are modified to suit their function. Some examples are shown in Figure 1.3.

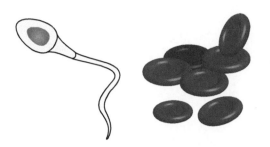

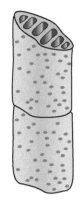

Sperm cell
The cell has very little cytoplasm and a tail, to help it swim fast towards the egg

Red blood cells
The cells have lost their nuclei and have become packed with a red pigment, haemoglobin, which carries oxygen around the body

Xylem cells
The xylem cells form tubes which carry water up a plant, and also strengthen it. To do this, the cells have perforated end walls, the cell wall is very thick, and the cytoplasm has died off to leave a hollow tubet

Figure 1.3 Specialised cells.

The table below shows how cells are gradually built up into a whole living organism, through several 'levels of organisation'.

Table 1.1 Levels of organisation in the structure of living things.

Level of organisation	Definition	Examples
Tissue	A group of similar cells with similar functions	Bone, muscle, blood, xylem, epidermis
Organ	A collection of two or more tissues that perform specific functions	Kidney, brain, heart, leaf, flower
Organ system	A collection of several organs that work together	Digestive system, nervous system, respiratory system, shoot system, root system
Organism	A whole animal or plant	Cat, elephant, human, rose bush, oak tree

Exam tip

In distinguishing plant and animal cells, the cell wall is the key feature, because some plant cells do not have chloroplasts and some animal cells have (temporary) vacuoles.

Diffusion, osmosis and active transport

The cell membrane is **selectively permeable**, and will allow some substances to pass through but not others. The movement through the membrane is by one of three processes – **diffusion, osmosis** or **active transport**.

Diffusion

Diffusion is the spreading of particles from an area of higher concentration to an area of lower concentration, as a result of random movement. We say the particles move down a **concentration gradient**. Diffusion is a **passive process** (it does not require energy). The particles will move in all directions, yet the *overall* (net) movement is from an area of high concentration to an area of low concentration. The speed of diffusion is affected by two factors:

1 Temperature – increasing the temperature will speed up diffusion, because it makes the particles move faster.
2 The size of the concentration gradient. The bigger the concentration gradient, the faster diffusion will be.

Osmosis

Osmosis is a specific type of diffusion. It is the diffusion of **water molecules** through a **selectively permeable membrane**. The water moves from a solution of low solute concentration (which has more water) to a solution of high solute concentration (which has less water). As with diffusion, water molecules move in both directions, but the **net movement** is from the more dilute to the more concentrated solution. The process of osmosis is summarised in Figure 1.4.

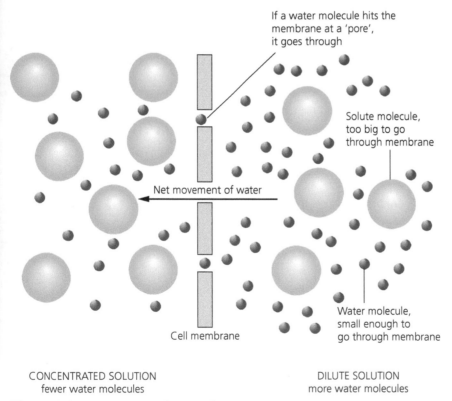

Figure 1.4 The process of osmosis.

If an animal cell is put into a solution that is more dilute than its cytoplasm, water will go in by osmosis and the cell will burst. Plant cells do not burst in dilute solutions because their cell wall prevents it.

Very concentrated solutions damage both plant and animal cells. They will lose water, and animal cells shrivel. In plant cells the cytoplasm shrinks and pulls away from the cell wall, a process known as **plasmolysis**.

Active transport

When cells need to move substances against a concentration gradient (from an area of lower concentration to an area of higher concentration), this will not happen by diffusion, and the cell has to use energy to 'pump' the particles in the direction they need to go. This type of transport requires energy, and is called **active transport**.

Now test yourself

1 Which three structures are only found in plant cells, never in animal cells?
2 What is the function of mitochondria in cells?
3 Which lens in a microscope is changed in order to alter the magnification?
4 State two differences between diffusion and active transport.
5 Why do animal cells burst when put in water, but plant cells do not?

Answers online

Enzymes

All the chemical reactions in the body are controlled by special molecules called **enzymes**. Which enzymes are produced in cells is controlled by another molecule, **deoxyribonucleic acid (DNA)**, which is found in the cell nucleus.

Properties of enzymes

- All enzymes are proteins.
- Different enzymes are made up of different sequences of amino acids. The bonds formed between the amino acids means that each type of enzyme has a specific molecular shape.
- Enzymes are **catalysts** – they speed up chemical reactions, without taking part in them.
- Enzymes are **specific**; they will only catalyse one reaction or one type of reaction.
- The rate of an enzyme-controlled reaction is affected by temperature, pH and the concentrations of the enzyme and the **substrate** (the chemical on which the enzyme works).
- High temperatures and extreme pH can alter the enzyme's structure so that it becomes **denatured** and can no longer work.

The lock and key theory

To catalyse a reaction, the enzyme molecule must collide with, and 'lock together' with its substrate. The place where the substrate fits into the enzyme is called the **active site**. The shapes of the enzyme's active site and of the substrate must match, so that they fit together like a lock and key. That is why enzymes are specific – they can only work with substances that fit into their active site.

The action of this 'lock and key' model is shown in Figure 1.5 overleaf.

> **Enzyme-substrate complex:** The structure formed when an enzyme and its substrate are fitted together.

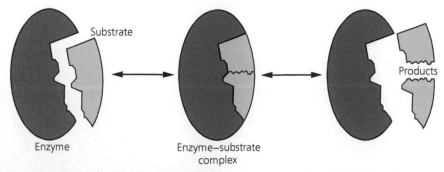

Substrate

Enzyme

Enzyme–substrate complex

Products

Figure 1.5 The 'lock and key' model of enzyme action. Note that in some reactions an enzyme catalyses the breakdown of a substrate into two or more products, while in others an enzyme causes two or more substrate molecules to join to make one product molecule.

The effect of temperature on enzymes

- Increasing temperature tends to speed up enzyme-controlled reactions, because it makes the particles move faster so they are more likely to collide.
- High temperatures break chemical bonds and change the shape of the enzyme molecule so that the substrate can no longer fit into the active site. The enzyme is denatured and cannot work.
- Denaturation often begins when temperatures go above 40 °C, although some enzymes can survive much higher temperatures.

The effect of temperature on enzymes is shown in Figure 1.6.

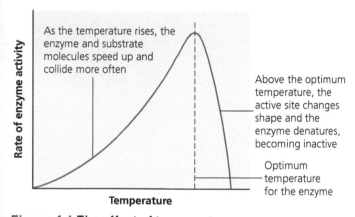

Rate of enzyme activity

As the temperature rises, the enzyme and substrate molecules speed up and collide more often

Above the optimum temperature, the active site changes shape and the enzyme denatures, becoming inactive

Optimum temperature for the enzyme

Temperature

Figure 1.6 The effect of temperature on enzymes.

> **Exam tip**
>
> Remember that enzymes do not take part in the reaction they catalyse. Never say that enzymes **react with** their substrate.

The effect of pH on enzymes

- Enzymes have an optimum pH, which varies from enzyme to enzyme.
- The further away the pH value is from the optimum, the less well the enzyme works.
- pH values that are a long way from the optimum can cause the enzyme to denature.

The effect of pH on enzymes is shown in Figure 1.7.

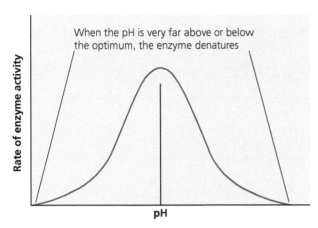

Figure 1.7 The effect of pH on enzymes.

Now test yourself

TESTED ☐

6 To what group of chemicals do enzymes belong?
7 What name is given to the chemical which an enzyme works on?
8 What is it about their structure that causes enzymes to be specific?
9 Why do high temperatures denature enzymes?
10 If you were investigating the effect of temperature on an enzyme, what other factors would need to be controlled?

Answers online

Summary

- Animal and plant cells have the following parts: cell membrane, cytoplasm, nucleus, mitochondria; in addition, plants cells have a cell wall, vacuole and sometimes chloroplasts.
- Cells differentiate in multicellular organisms to become specialised cells, adapted for specific functions.
- Tissues are groups of similar cells with a similar function; organs may comprise several tissues performing specific functions; organs are organised into organ systems, which work together in organisms.
- Diffusion is the passive movement of substances, down a concentration gradient.
- The cell membrane forms a selectively permeable barrier, allowing only certain substances to pass through.
- Osmosis is the diffusion of water through a selectively permeable membrane from a region of high water (low solute) concentration to a region of low water (high solute) concentration.

- ℍ *Active transport is an active process by which substances can enter cells against a concentration gradient.*
- Enzymes control the chemical reactions in cells; they are proteins made by living cells, which speed up – or catalyse – the rate of chemical reactions.
- The specific shape of an enzyme enables it to function, the shape of the active site allowing it to bind to its appropriate substrate.
- Enzyme activity requires molecular collisions between the substrate and the enzyme's active site.
- Increasing temperature increases the rate of enzyme activity, up to an optimum level, after which any further increase results in the enzyme being denatured. Boiling denatures most enzymes.
- Enzyme activity varies with pH. For each enzyme, there is an optimum pH, which is different for different enzymes.

Exam practice

1 Animals need to maintain a fairly constant level of salts in their bodies. Marine birds take in a lot of salt water when they feed on fish – too much for their kidneys to remove. They have special salt glands near their eyes which constantly remove salt from the blood and then discharge a strong salt solution out via the nostrils. The concentration of salt in the cells of the salt gland is always much higher than in the blood.

   a) What process would the cells use to absorb the salt from the blood? [1]
   b) Suggest a reason for your answer to (a). [1]
   c) The cells of the salt glands contain many mitochondria. Suggest a reason for this. [2]
   d) Birds have a waterproof skin. If this were not the case, what would happen to their body cells when they swam in sea water which is more concentrated than their body cells? [2]

2 Cylinders of equal length and diameter were cut from a potato, weighed and placed in different concentrations of salt solution for 30 minutes. They were then re-weighed. The table below shows the results.

Concentration of salt solution (M)	Initial mass (g)	Mass after 30 minutes (g)	Change in mass (g)	Percentage change in mass
0 (water)	4.1	4.7	+0.6	+14.6
0.2	3.4	3.6	+0.2	+5.9
0.4	4.0	3.5	−0.5	−12.5
0.6	4.5	3.7	−0.8	−17.8
0.8	4.2	3.1	−1.1	−26.2
1.0	3.7	2.6	−1.1	−29.7

   a) What process caused the potato cylinders to change in mass? [1]
   b) Calculate the percentage change in mass for the potato in 0.8 M salt solution. [2]
   c) Explain the difference between the results for 0–0.2 M salt solution and those for 0.4–1.0 M salt solution. [5]
   d) Approximately what concentration of salt solution is equal to the concentration of the potato cell sap? [1]
   e) Why was it necessary to record the **percentage** change in mass, and not just the change in mass? [1]

Answers and quick quiz 1 online

ONLINE

2 Respiration and the respiratory system in humans

Aerobic and anaerobic respiration

Aerobic respiration

Respiration is the process in every living cell by which food is broken down and the energy in it is released for use. The energy is temporarily stored and moved around the cell in the form of a chemical called **adenosine triphosphate (ATP)**.

The usual food molecule respired is glucose (although it is possible to use others) and oxygen is used up in the process if respiration is aerobic, which it usually is. Carbon dioxide and water are produced as waste materials from the process. The word equation summarising aerobic respiration is:

Glucose + oxygen → carbon dioxide + water + ENERGY

Aerobic respiration is actually a complex series of chemical reactions, each one controlled by a different enzyme.

Anaerobic respiration

When oxygen is in short supply or absent, some types of cell can respire anaerobically (without oxygen). During periods of intense activity, human muscles sometimes need more oxygen than the body can supply. The muscle cells use **anaerobic respiration** to partially breakdown glucose and release some of the energy from it. The equation for anaerobic respiration is:

Glucose → lactic acid + ENERGY

Anaerobic respiration is a much less efficient process than aerobic respiration, because the glucose is not fully broken down and so much less ATP is formed for each molecule of glucose used. For that reason, animal cells always respire aerobically when they can. Some microorganisms respire anaerobically all the time, however.

Oxygen debt

During vigorous exercise, breathing cannot supply the muscles with all the oxygen they need, and so they switch to anaerobic respiration. **Lactic acid** builds up, which causes the muscles to ache.

Oxygen breaks down lactic acid and releases the remaining energy. So, when you finish the exercise, your body keeps breathing faster and deeper to provide extra oxygen to break down the lactic acid. In effect, you are breathing in the oxygen that you needed (but could not get) during the exercise. You have built up an **oxygen debt**, which is then repaid after the exercise is finished.

Respiration and breathing

Respiration is not the same thing as breathing.

Respiration goes on in all living cells, releasing energy from food. In general, oxygen is needed for this.

Breathing is the way some animals get the oxygen they need for respiration. Plants do not breathe, and in fact many animals do not either. Many small animals can absorb oxygen through the surface of their body.

The composition of the air we breathe in and out is shown in Table 2.1

Table 2.1 Approximate composition of inspired and expired air.

Gas	% inspired air	% in expired air
Oxygen	21	16
Carbon dioxide	0.04	4
Nitrogen	79	79

Now test yourself

1 Which food chemical is most often used for respiration?
2 Give three differences between aerobic and anaerobic respiration.
3 Which human cells are capable of anaerobic respiration?
4 Why would cells be unable to respire at very high temperatures?
5 After exercise, how does a person 'pay back' their oxygen debt?

Answers online

The respiratory system and breathing 1

The need for a respiratory system

Very small animals exchange gases by diffusion across their body surface, but larger animals need a respiratory system. There are several reasons for this:
- Larger animals have a smaller surface area: volume ratio. The surface area is insufficient to supply the volume with oxygen.
- Diffusion is very slow. The innermost cells in a large animal would die due to a lack of oxygen before oxygen from the surface could reach them.
- Larger animals are usually more active than very small ones, so more oxygen is needed for extra energy.

Structure of the respiratory system

The structure of the respiratory system is shown in Figure 2.1

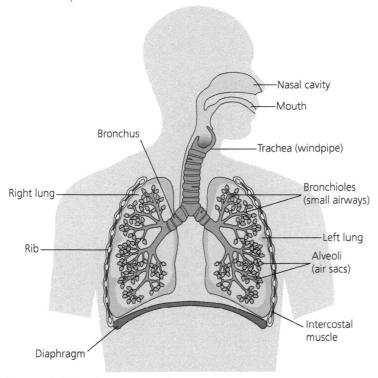

Figure 2.1 The human respiratory system.

The parts of the respiratory system are as follows:
- The **trachea** is the tube, also known as the windpipe, which leads from the mouth and nasal cavity towards the lungs. It is strengthened by rings of cartilage.
- The trachea branches into two **bronchi** (singular: **bronchus**), each of which leads to one of the lungs. These also have rings of cartilage.
- The bronchi branch into many **bronchioles** which spread throughout the lungs. The bronchioles to not have rings of cartilage.
- At the end of each bronchiole are clusters of air sacs called **alveoli** (singular: **alveolus**). This is where oxygen enters the blood and carbon dioxide is removed.
- The **lung** is the name given to the organ made up of the bronchioles, alveoli and the tissues surrounding them.
- The **ribs** surround the lungs. They protect them and also help with breathing, due to the action of the intercostal muscles between them.
- The **diaphragm** is a sheet of muscle below the lungs. It also helps with breathing.

The breathing mechanism

The mechanism for breathing in (inspiration) and out (expiration) is shown in Figure 2.2 and Table 2.2.

H It relies on the fact that air will always move from areas of higher pressure to areas of lower pressure.

Table 2.2 Mechanism for inspiration and expiration.

Inspiration	Expiration
The ribs move up and out	The ribs move down and in
The diaphragm flattens (moves downwards)	The diaphragm bows upwards
The volume of the thorax increases and the pressure decreases	The volume of the thorax decreases and the pressure increases
The lungs expand	The lungs recoil back to their original shape
The pressure in the lungs is lower than the outside air, so air is sucked in through the trachea	The pressure in the lungs is higher than outside and so air moves out through the trachea

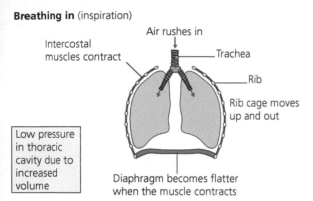
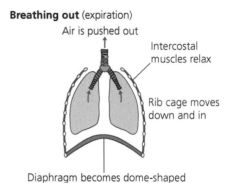

Figure 2.2 Mechanism for breathing in and out.

The respiratory system and breathing 2

Adaptations of the respiratory system for efficient gas exchange

REVISED

The **alveoli** are the gas exchange surface. They have the following adaptations for efficiency:
- They have a very large surface area.
- They are well supplied with blood vessels, to take the absorbed oxygen away.
- The surface of the alveoli is only one cell thick, so the gases have a very small distance to diffuse.
- The surface is moist, to dissolve the oxygen so that it can diffuse through the membrane.

Preventing lung infections

REVISED

The lungs are protected against dust and microorganisms which enter with the air that is breathed in. The cells lining these tubes produce **mucus**, which is a sticky substance that traps dust and microbes from the air as it passes through. The cells lining the trachea and bronchi have small, hair-like structures on them called **cilia**. These constantly move back and forth, pushing the mucus up towards the top of the trachea, where it can be swallowed and eliminated from the body via the digestive system.

> **Exam tip**
>
> When breathing out, the lungs do reduce in volume, but do not say they 'contract'. Contract is a term that is linked with muscles, and the change in volume of the lungs is caused by elastic recoil, not muscles.

Smoking and the lungs

Smoking is known to damage the lungs in several ways. The chemicals in tobacco smoke that do the damage are as follows:

- **Carcinogens** – chemicals that cause cancer (43 different substances).
- **Tar** – a sticky substance which clogs the bronchioles and alveoli.
- **Nicotine** – the main problem with nicotine is that it is extremely addictive, although it also directly damages the lungs.
- **Carbon monoxide** – a poisonous gas which makes it more difficult for the red blood cells to carry oxygen.
- There are several other harmful substances in low quantities, e.g. ammonia, formaldehyde, hydrogen cyanide and arsenic.

The effects of smoke on the lungs can be:

- Lung cancer
- Other cancers (e.g. of the mouth, oesophagus, bladder, kidney and pancreas)
- Emphysema (damage to the walls of the alveoli)
- Paralysis (and eventual destruction) of the cilia lining the respiratory system. This means that the mucus with its trapped microorganisms and dust sinks into the lungs, rather than being moved up the trachea to be swallowed.

Now test yourself

TESTED

6 Why is it important that the walls of the alveoli are very thin?
7 Which structures cause the ribs to move during breathing?
8 When the lungs expand, why does air move into them from the outside?
9 What role does mucus play in the protection of the lungs?
10 Which chemical in tobacco makes smoking addictive?

Answers online

Summary

- Aerobic respiration is a series of enzyme-controlled reactions that occur in cells when oxygen is available.
- Aerobic respiration uses glucose and oxygen to release energy* and produces carbon dioxide and water.
  *(in the form of ATP)
- Anaerobic respiration occurs when oxygen is not available. In animals, glucose is broken down into lactic acid.
- Anaerobic respiration in muscles builds up an oxygen debt, which is repaid after the exercise by breathing faster and deeper than normal.
- Anaerobic respiration produces less ATP (per molecule of glucose) than aerobic.
- Larger animals need a respiratory system because diffusion over the surface cannot supply the increased volume of the organism with oxygen, and diffusion is too slow to reach the centre of the organism.
- The respiratory system consists of the following structures: nasal cavity, trachea,

bronchi, bronchioles, alveoli, lungs, diaphragm, ribs and intercostal muscles.
- Mucus lining the respiratory system traps dust and microbes. The cilia on the cells of the breathing tubes move the mucus to the top of the trachea, where it can be swallowed.
- The cilia are paralysed by tobacco smoke, so that the mucus sinks into the lungs, carrying the dust and microbes with it.
- Movements of the ribs and diaphragm cause breathing in (inspiration) and breathing out (expiration).
- Movement of air takes place due to differences in pressure between the lungs and the outside of the body.
- Gas exchange occurs at the alveoli, which have thin walls, a moist lining and a good blood supply.
- Expired air contains more carbon dioxide and less oxygen than inspired air.
- Smoking is a major contributory factor in lung cancer and emphysema.

Exam practice

1 100 m sprinters' muscles respire anaerobically for all except the first few metres of a race. The muscles of a long-distance runner rarely respire anaerobically during a race.
   a) Write a word equation for anaerobic respiration. [1]
   b) What causes a sprinter's muscles to respire anaerobically? [3]
   c) Why would it be a problem for long-distance runners if their muscles respired anaerobically for a considerable time? [2]
   d) Why do sprinters breathe more rapidly and deeply for a short period after their race finishes? [2]

2 The diagram shows an experiment to measure the rate of respiration of germinating peas, using a piece of apparatus called a respirometer. When the peas absorb oxygen, the volume of air in the respirometer decreases and this pulls the liquid along the capillary tube. The experimenter measures how far the liquid moves, at two minute intervals. The respirometer with the glass beads is a control experiment.

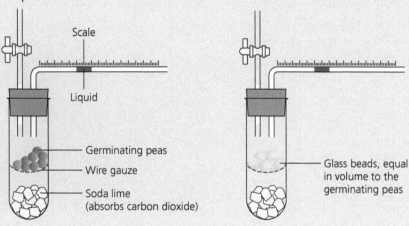

   a) Why is it necessary to have soda lime in the respirometer to absorb carbon dioxide? [3]
   b) What is the purpose of the glass beads in the control experiment? [1]
   c) The results are shown in the table. Calculate the average rate of movement of the liquid, in mm min⁻¹. [2]
   d) A student suggested that the experiment could be improved by putting the respirometers in a water bath at 30 °C. Suggest a reason for this. [2]

Time (minutes)	Distance moved (mm)
2	20.4
4	41.7
6	75.3
8	95.5
10	103.8

Answers and quick quiz 2 online

ONLINE

3 Digestion and the digestive system in humans

The process of digestion

The need for digestion
REVISED

The food we eat contains large and complex molecules which cannot get through the gut wall and into the blood. They need to be digested, to accomplish two things:
- Large molecules must be broken down into smaller molecules that can get through the gut wall.
- Insoluble molecules must be converted into soluble ones, so that they can be transported around the body in the blood.

The end point of digestion is small, soluble molecules. Some food molecules are already small and soluble (e.g. glucose, vitamins) and do not need digesting.

Digestion of different food molecules

- Large carbohydrates are digested into simple sugars. In the case of starch, the simple sugar produced is glucose.
- Fats are broken down into **glycerol** and **fatty acids**.
- Proteins are broken down into **amino acids**.

Food tests
REVISED

Different food groups can be identified in samples by a variety of chemical tests.

Starch

Add **iodine** solution. A colour change from brown to blue-black indicates the presence of starch.

Glucose

Add **Benedict's solution** and heat in a boiling water bath. If glucose is present, the solution goes cloudy and changes from blue to green then orange and finally brick red.

Protein

Add **Biuret** solution (sometimes this is added as two separate parts – dilute sodium hydroxide solution followed by dilute copper sulfate solution). A colour change from pale blue to purple indicates the presence of protein.

> **Exam tip**
>
> When describing the colour changes in a food test, always give the starting colour as well as the end colour, as there may be a mark for it.

The digestive system

The digestive system (gut) is a tube that goes through the body, with some additional organs which are associated with it. As it goes through the gut the food goes through three processes:

- **Digestion** – the breakdown of food, mainly in the mouth, stomach and small intestine.
- **Absorption** – the transfer of the breakdown products into the blood, mainly in the small intestine (food) and the large intestine (water).
- **Egestion** – the removal of any undigested food via the rectum and the anus.

Different parts of the system are specialised for different functions, as shown in Figure 3.1.

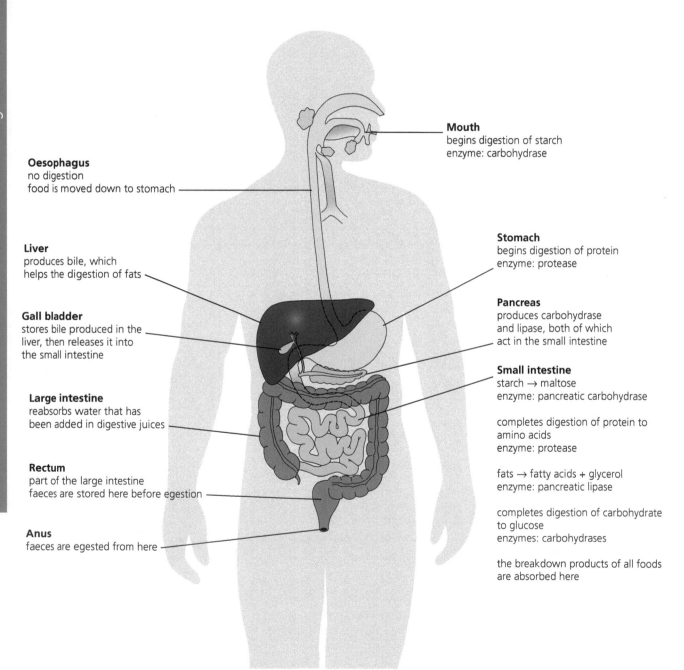

Oesophagus
no digestion
food is moved down to stomach

Liver
produces bile, which
helps the digestion of fats

Gall bladder
stores bile produced in the
liver, then releases it into
the small intestine

Large intestine
reabsorbs water that has
been added in digestive juices

Rectum
part of the large intestine
faeces are stored here before egestion

Anus
faeces are egested from here

Mouth
begins digestion of starch
enzyme: carbohydrase

Stomach
begins digestion of protein
enzyme: protease

Pancreas
produces carbohydrase
and lipase, both of which
act in the small intestine

Small intestine
starch → maltose
enzyme: pancreatic carbohydrase

completes digestion of protein to
amino acids
enzyme: protease

fats → fatty acids + glycerol
enzyme: pancreatic lipase

completes digestion of carbohydrate
to glucose
enzymes: carbohydrases

the breakdown products of all foods
are absorbed here

Figure 3.1 The human digestive system.

Peristalsis

Food is moved through the digestive system by a process called peristalsis. This involves waves of muscle contractions in the gut wall (behind the food) which squeezes it along the gut. It is shown in Figure 3.2.

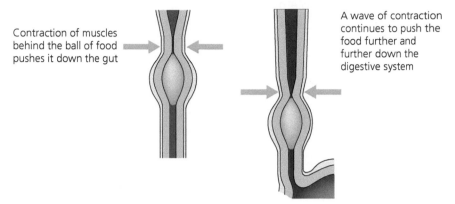

Contraction of muscles behind the ball of food pushes it down the gut

A wave of contraction continues to push the food further and further down the digestive system

Figure 3.2 Peristalsis in the gut.

Digestive enzymes

The reactions that digest food are all catalysed by enzymes. These are of three main types:

- **Carbohydrases** – these breakdown complex carbohydrates into simple sugars like glucose. They are produced in the mouth, the pancreas and the small intestine.
- **Proteases** – these digest proteins into amino acids, and are found in the stomach and the small intestine.
- **Lipases** – these convert fats into glycerol and fatty acids. They are produced by the pancreas and the small intestine.

The role of the different parts of the digestive system and the enzymes they contain are shown in Table 3.1.

Table 3.1 Digestive system and enzymes.

Part of the digestive system	Changes taking place	Enzymes	Action of enzymes
Mouth	Food is lubricated and swallowed. Some starch is digested	Carbohydrase (amylase)	Starch › maltose
Oesophagus	None	None	
Stomach	Hydrochloric acid is added which kills bacteria. Some proteins are digested	Protease	Proteins → peptides
Pancreas	Enzymes delivered to the small intestine, where they act	Carbohydrase (amylase)	Starch → maltose
		Lipase	Fats → Fatty acids + glycerol
Small intestine	Digestion is completed. Enzymes produced in the pancreas act here, as well as those produced by the small intestine. After digestion, the products are absorbed into the blood	Carbohydrases	Complete carbohydrate digestion to simple sugars
		Proteases	Complete protein digestion to amino acids
Large intestine	Most of the water is reabsorbed	None	
Rectum and anus	Waste is egested	None	

Bile

Bile is a liquid produced in the liver and stored in the gall bladder that helps digestion, even though it contains no enzymes. Bile travels down the bile duct to the small intestine, where it helps lipase enzymes to digest fats. Bile **emulsifies** the fats, splitting them into small droplets, and this provides a greater surface area for the lipase enzymes to work on (see Figure 3.3).

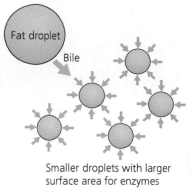

Fat droplet

Bile

Smaller droplets with larger surface area for enzymes

Figure 3.3 The effect of bile on fats.

The absorption of food, and diet

Absorption and egestion

Digested food is absorbed in the second half of the small intestine. To help this, the walls of the small intestine are covered in small, finger-like projections called **villi**, which greatly increase the surface area over which food can be absorbed. The structure of a villus is shown in Figure 3.4. By the time the gut contents reach the large intestine, all of the useful products have been absorbed into the blood. The large intestine reabsorbs most of the water from the waste, which therefore solidifies into **faeces** as it goes down the large intestine. The faeces are temporarily stored in the rectum before being egested via the anus.

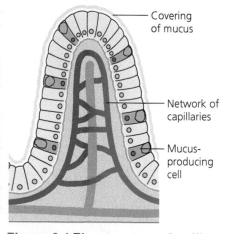

Covering of mucus

Network of capillaries

Mucus-producing cell

Figure 3.4 The structure of a villus.

Now test yourself

TESTED

1 What chemicals are produced by the complete digestion of fats?
2 Which parts of the digestive system contain protease enzymes?
3 What is the function of the large intestine?
4 Where is bile stored?
5 How does bile help the digestion of fats?

Answers online

A balanced diet

Humans require a variety of nutrients, as each type has a different function in the body.

- **Glucose**, formed by the breakdown of carbohydrates, is the main energy provider in the body. It is stored as **glycogen** in the liver.
- **Fatty acids and glycerol** from fats also provide energy. Fats contain more energy per gram than glucose, but it can only be released slowly. For this reason, fats are useful as an energy store.
- **Amino acids** from proteins are re-assembled in the body into new proteins, to form many useful products or to be used for making new cells in growth.

Apart from the chemicals that are formed by the digestion of food, there are other useful substances in the diet that can be absorbed directly because they are small molecules.

- **Minerals** have a variety of functions – e.g. iron is needed to make haemoglobin, the blood pigment that carries oxygen around the body.
- **Vitamins** also perform various jobs, and are often needed for important chemical reactions to take place. Vitamin C, for example, helps the immune system to function properly.
- **Water** is important because it is the main constituent of cells, and all of the chemical reactions in the body involve chemicals that must be dissolved in water.

Finally, there are health benefits from having quite a lot of **fibre** in the diet. Fibre is indigestible but it provides bulk for the gut to act on during peristalsis, and so aids the efficient movement of food through the gut.

> **Exam tip**
>
> The term 'nutrients' relates to things that provide nourishment for the body. Water and fibre do not come in that category so, if a question asks about nutrients, stick to carbohydrates, proteins, fats, minerals and vitamins.

How diet affects health

Different foods contain different amounts of energy, but the main type of food the body uses for energy is **glucose**, a sugar that we get from eating any **carbohydrate**. Fats contain more energy than carbohydrates, but they are used more slowly and are generally used only when carbohydrate levels are low. Protein is not normally used for energy as the body cannot store it. If we eat more carbohydrate than we need at the time, the body stores it in the liver for future use as a substance called **glycogen**. If we keep eating more food than we need, this store becomes full. The body then changes the carbohydrate into **fat**, which is stored under the skin and around the internal organs.

In other words, we 'get fat'. These fat stores also increase directly if excess fat is eaten. If we eat less and exercise more, these stores get used up. Anyone who is severely overweight has an increased risk of several serious conditions, e.g.

- Heart disease
- Stroke
- Cancers
- Type 2 diabetes

Sugar is the worst form of carbohydrate to eat as it is very easily digested and absorbed. The body cannot store protein so eating a lot of protein has little effect on health. Sugar is added to many processed foods for taste and also as a preservative.

Excess salt can also be a hazard, as it can lead to high blood pressure which increases the risk of heart disease and strokes. Salt is a common food additive.

Visking tubing as a model gut

Rather than using pieces of gut when doing experiments about digestion, scientists often use a type of soft plastic tubing called **Visking tubing**. It is a good, but not perfect, model of the lining of the gut. Its advantage is that it has similar permeability to the lining of the intestine and will let the same type of chemicals through. Its disadvantage is that it is non-living. Living cells are capable of pumping substances through their cells by active transport, using energy from respiration. This cannot happen with Visking tubing. The Visking tubing has small holes (pores) in it which let small molecules through. The gut lining does not have holes in it.

Now test yourself

6 Which type of food provides the most energy per gram?
7 Why is it particularly important that babies have sufficient protein in their diet?
8 What is the function of fibre in the diet?
9 List three conditions where the risk of having the condition is greater in overweight people.
10 Explain why exercise reduces the risk of becoming overweight.

Answers online

Summary

- Complex, insoluble food molecules are broken down into small, soluble molecules that can enter the blood system. This breakdown, in the digestive system, is called digestion.
- Digestion is aided by enzymes.
- Visking tubing behaves similarly to the wall of the gut, and it can be used as a 'model gut'.
- Fats are digested into fatty acids and glycerol.
- Proteins are digested into amino acids.
- Starch is digested into glucose.
- The test for starch uses brown iodine solution, which turns blue-black.
- The test for glucose is the Benedict's test. The test sample is boiled with Benedict's solution (blue) and, if glucose is present, a reddish-orange precipitate is formed.
- The test for protein is the Biuret test. Copper sulfate and sodium hydroxide are added to the test solution. If protein is present, a purple colour appears.
- The digestive system consists of the mouth, oesophagus, stomach, small intestine, large intestine, anus, liver, gall bladder and pancreas.
- The mouth contains carbohydrase, which digests starch.
- The stomach contains protease, which digests proteins.
- The small intestine contains various enzymes, which complete the digestion of carbohydrates, proteins and fats.

- The liver produces bile, which is stored in, and released from, the gall bladder. Bile emulsifies fats, which aids their digestion.
- Food is moved along the digestive system by peristalsis.
- Glucose from carbohydrates, and fatty acids and glycerol from fats, provide energy for the body.
- Amino acids from proteins form the building blocks for new proteins, which are needed for growth and repair of tissues and organs.
- For optimum health, we need to eat a balanced diet, with appropriate levels of carbohydrates, fats, proteins, minerals, vitamins, water and fibre.
- Carbohydrates, fats and proteins all contain energy.
- Fats contain the most, with carbohydrates and proteins having less (and roughly equal) amounts.
- Our bodies use carbohydrates and fats for energy.
- If we take in too much fat or carbohydrate, the extra energy is stored as fat.
- A diet that is too high in sugar or fat can lead to health problems.
- Additives are added to our food. Salt is a common one, and eating too much salt can lead to high blood pressure and associated health problems.

Exam practice

1 The diagram shows an experiment to investigate the digestion of starch by amylase (a carbohydrase enzyme) using Visking tubing as a model of the gut lining.

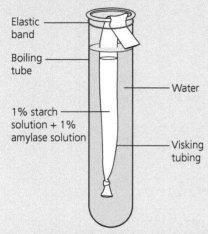

Figure 3.5 Apparatus used to model absorption in the gut.

After a few hours, the liquid in the tube (outside the Visking tubing) was tested for starch and glucose. The results are shown in the table below:

Test	Colour seen
Iodine test	Brown
Benedict's test	Brick red

a) What do you conclude from the observations? [2]
b) Describe how the Benedict's test is carried out. [2]
c) Explain the Benedict's test result. [3]
d) A student suggested that heating the apparatus would make the results detectable earlier. Suggest a possible problem that might occur if high temperatures were used. [2]
e) Apart from raising the temperature, suggest one other way that results might be obtained more quickly. [1]

2 The table below gives the energy requirement of human males at different ages:

Age (years)	Energy requirement per day (kcal/day)
5	1482
10	2032
15	2820
19–24	2772
45–54	2581
65–74	2294

a) Suggest a reason why the energy requirement goes down after age 15? [2]
b) The mean weight of males between 19–24 is 76 kg. Calculate their energy requirement per kg of body weight. [2]
c) The mean energy requirement of 15-year-old girls is 2390 kcal/day. Suggest why this figure is lower than for boys. [2]
d) When comparing the energy requirements of 5 year olds and adults, suggest a reason why it would be better to use kilocalories per kg of body weight per day, rather than just kcal/day. [1]

Answers and quick quiz 3 online

ONLINE

Blood and the circulatory system

The human circulatory system transports materials around the body. It is filled with a liquid (blood) which can carry all the essential materials. The blood is pumped around the system by the heart, through a series of blood vessels.

Substances transported by the blood REVISED ☐

- **Digested food** is transported from the digestive system to all parts of the body, in the blood plasma.
- **Oxygen** is transported from the lungs to the body in the red blood cells.
- **Carbon dioxide**, produced in respiration is transported to the lungs, where it is breathed out.
- **Wastes** are transported to the kidneys for excretion.
- **Hormones** (chemical messengers which regulate processes in the body) are transported in the plasma.

The blood is also important in fighting **pathogens** (microorganisms which cause disease) and in maintaining body temperature.

Blood REVISED ☐

Blood is made up of the following parts:
- **Plasma**. The liquid part of the blood which transports water soluble substances (e.g. digested food, carbon dioxide, urea, salts and hormones).
- **Red blood cells**. These are responsible for the transport of oxygen, attached to the red pigment **haemoglobin**.
- **White blood cells**. These cells are not involved in transport, but fight diseases as part of the immune system.
- **Platelets**. These are cell fragments that help the blood to clot.

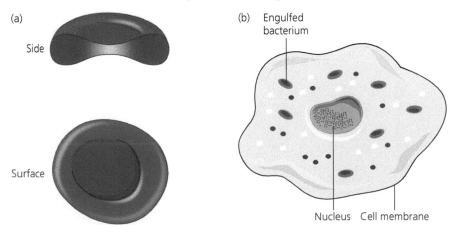

Figure 4.1 Structure of (a) a red blood cell and (b) a phagocyte (a type of white blood cell).

Figure 4.1 shows the structure of red and white blood cells. The essential features of the two types of blood cell are listed.

Red blood cells

- They are **biconcave discs** – round and flattened, with a central indentation. This shape increases the surface area for absorption of oxygen, compared with a rounded shape.
- They contain the red blood pigment **haemoglobin**, which absorbs oxygen.
- They have lost their nucleus, which allows more haemoglobin to be packed into their cytoplasm.

White blood cells

- They can change shape and can also move. This allows them to ingest bacteria and squeeze through tiny gaps in capillary walls to enter the tissue fluid to fight infection.
- There are different types of white blood cell which do different jobs. Some (**phagocytes**) ingest bacteria, others are involved in the production of antibodies.

The circulatory system

REVISED

The blood is pumped through the circulatory system by the heart, and moves around the body in blood vessels called **arteries**, **veins** and **capillaries**. It travels through the blood vessels in the following sequence:

heart → arteries → capillaries → veins → heart

A simplified structure of the circulatory system of a mammal is shown in Figure 4.2.

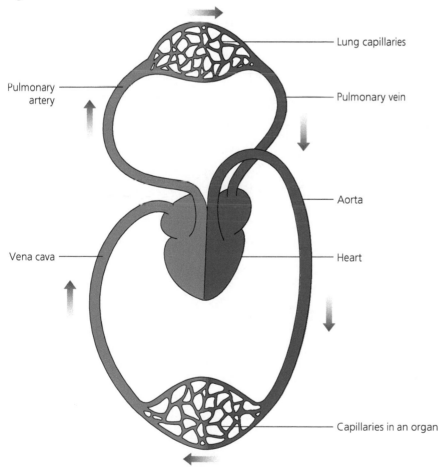

Figure 4.2 Structure of the circulatory system of a mammal. The arrows show the direction of blood flow.

The circulatory system in mammals is referred to as a **double circulation**, because, in any one complete circuit of the body, the blood travels through the heart twice. The heart is divided into two halves:

- The left half receives **oxygenated blood** from the lungs, and pumps it to the rest of the body.
- The right half receives **de-oxygenated blood** from the body and pumps it back to the lungs.

Figure 4.2 shows the two circulations.

- The **pulmonary circulation** takes blood to and from the lungs.
- The **systemic circulation** takes blood to and from the body.

Oxgenated blood: blood with a high level of oxygen in it.

De-oxygenated blood: blood which has had most of the oxygen removed by the tissues.

Now test yourself

TESTED

1 Name the two important gases which are transported by the blood.
2 Which part of the blood helps with clotting?
3 How does the lack of a nucleus help red blood cells perform their function?
4 Which type of blood vessel carries blood away from the heart?
5 Which two organs does blood travel through in the pulmonary circulation?

Answers online

The heart

Blood is moved around the body by the pumping of the heart. The heart is made of muscle, which, when it contracts, applies a force to the blood and pushes it out into the arteries. The exterior of the heart is shown in Figure 4.3. The outside of the heart has its own blood supply, via the **coronary artery**. Even though the heart is filled with blood, the muscular walls are so thick that the outside needs a separate blood supply. The blood supplies the nutrients and oxygen that the heart needs to keep beating.

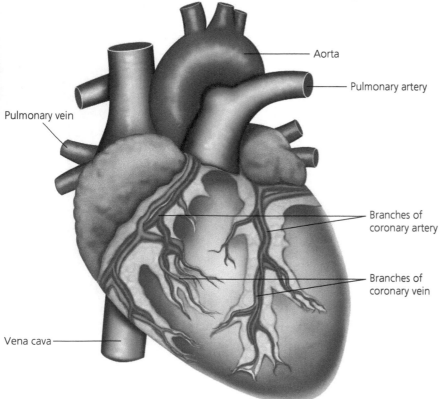

Aorta

Pulmonary artery

Pulmonary vein

Branches of coronary artery

Branches of coronary vein

Vena cava

Figure 4.3 Exterior view of the human heart.

Figure 4.4 shows the way in which blood flows through the heart. The right side of the heart deals with **de-oxygenated blood** and the left side with **oxygenated blood**. Another difference is that the left ventricle has a much thicker wall than the right ventricle, because it has to pump blood all around the body, whereas the right ventricle only has to pump blood to the lungs (which are very close to the heart). Notice that the right and left sides of the heart refer to the right and left of the person whose heart it is, not the viewer's right and left.

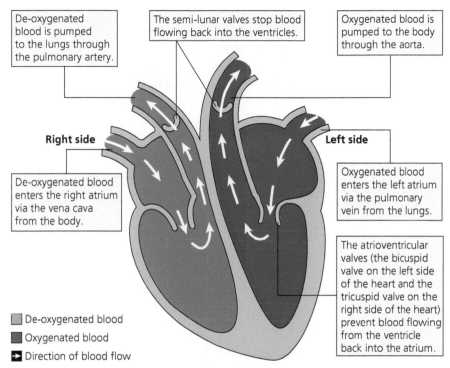

De-oxygenated blood is pumped to the lungs through the pulmonary artery.

The semi-lunar valves stop blood flowing back into the ventricles.

Oxygenated blood is pumped to the body through the aorta.

Right side

Left side

De-oxygenated blood enters the right atrium via the vena cava from the body.

Oxygenated blood enters the left atrium via the pulmonary vein from the lungs.

The atrioventricular valves (the bicuspid valve on the left side of the heart and the tricuspid valve on the right side of the heart) prevent blood flowing from the ventricle back into the atrium.

■ De-oxygenated blood
■ Oxygenated blood
➡ Direction of blood flow

Figure 4.4 Blood flow through the heart. Functions of the heart and blood vessels.

Valves and one-way flow in the heart

REVISED

Blood must flow through the heart in one direction only, from the **atria** to the **ventricles** and then out of the arteries at the top. Ensuring one-way flow is the job of the **valves**. The **atrioventricular** (bicuspid and tricuspid) valves between the atria and the ventricles stop backflow from the ventricles into the atria, and the **semilunar** valves at the beginning of the aorta and the pulmonary artery make sure that blood that has left the heart is not sucked back when the heart relaxes.

● Blood flows through the heart in the following way (the left side is used in this example).
● Oxygenated blood flows into the left atrium via the pulmonary vein from the lungs.
● The left atrium contracts, forcing the bicuspid valve open so that blood goes into the left ventricle.
● The left ventricle contracts, which forces the bicuspid valve shut but opens the semilunar valve.
● Blood flows out of the heart via the aorta.

Now test yourself

6 Which blood vessel supplies the outside of the heart with food and oxygen?
7 Which side of the heart deals with de-oxygenated blood?
8 Why does the left ventricle have a thicker wall than the right ventricle?
9 Which two blood vessels have semilunar valves at their base?
10 Which blood vessel brings blood back to the heart from the main part of the body?

Answers online

Arteries, veins and capillaries

There are three types of blood vessel in the body – **arteries, veins** and **capillaries**. Arteries carry blood away from the heart at high pressure, because the beating of the heart puts pressure on the blood.

The capillaries are very small and there are a very large number in each organ. It is in the capillaries that exchange of materials occurs. Oxygen and nutrients are delivered to the cells and waste products (including carbon dioxide) are picked up. Capillaries are very narrow and so the blood flows slowly through them. This, and the fact that there are so many of them, means that a lot of materials can be exchanged.

The capillaries eventually discharge their blood into veins, which take it back to the heart.

> **Exam tip**
>
> A common mistake in exams is to say that 'capillaries are one cell thick'. That is incorrect – it is their walls that are one cell thick.

Cardiovascular disease

Cardiovascular disease (CVD) is a common cause of death. CVD includes all the diseases of the heart and the circulatory system, including:
- Coronary heart disease
- Heart attacks
- Angina
- Strokes

CVD is usually linked to a process called **atherosclerosis** (Figure 4.5). This is the build-up of a substance called **plaque** in the walls of the arteries. Atherosclerosis has a number of effects.
- It makes it more difficult for blood to flow through the arteries, which means that the heart needs to work harder.
- It can block smaller arteries, depriving the tissues they supply of oxygen and nutrients. If this happens in the coronary artery, a **heart attack** can result.
- A blockage in the coronary artery can cause pains in the chest, a condition known as **angina**.
- The slower flow of blood makes it more likely that a clot will form, which can also block off blood vessels and cause a heart attack. If the blockage happens in the brain, a **stroke** can result.

> **Exam tip**
>
> Cardiovascular disease is not an alternative term for a heart attack. A heart attack is one (of several) forms of CVD.

Normal artery

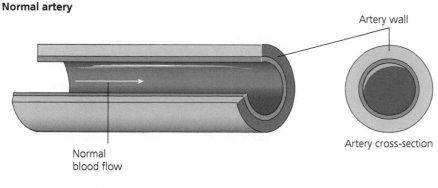

Narrowing of artery

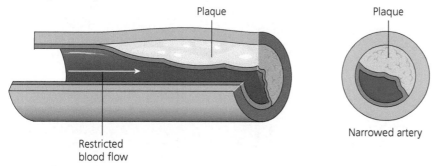

Figure 4.5 The process of atherosclerosis.

Reducing the risk of cardiovascular disease

The following are known risk factors for CVD. Some can be avoided by adopting a healthier lifestyle, but others cannot:

- **High blood pressure** – High blood pressure means that your heart is having to work harder than is ideal, which puts a strain on both your heart and your blood vessels. One cause of high blood pressure is a diet that is too high in salt.
- **Smoking** – Tobacco smoke contains carbon monoxide, which limits how much oxygen the blood can absorb. To get the right amount of oxygen to the tissues, the heart needs to work harder. Smoking can also lead to high blood cholesterol (see below).
- **High blood cholesterol** – One form of cholesterol is the substance that forms plaques in the artery walls, so a high-cholesterol diet can increase the likelihood of atherosclerosis. A high-cholesterol diet is one that contains too much saturated fat (which the body converts into cholesterol).
- **Diabetes** – Both type 1 and type 2 diabetes increase the risk of CVD.
- **Being overweight or obese** – Excess weight causes a build-up of fat around the organs including the heart. The heart has to work harder to provide the energy to move the extra bulk around.
- **Lack of exercise** – Exercise improves the condition of the heart and helps to avoid becoming overweight or obese.
- **A family history of heart disease** – Having close relatives with heart disease increases the risk of CVD, suggesting that a person's genes can make heart disease more, or less, likely.
- **Ethnic background** – People of a South Asian or African Caribbean ethnicity have a statistically higher risk of CVD than people from other ethnic backgrounds.

Now test yourself

11 Why do veins need valves?
12 Why do arteries have thick walls?
13 What is atherosclerosis?
14 What causes a stroke?
15 Why does a diet high in cholesterol increase the risk of CVD?

Answers online

Summary

- Blood has two functions in the body – transport and immunity.
- Blood consists of plasma, red cells, white cells and platelets.
- The red cells are responsible for the transport of oxygen, attached to the red pigment haemoglobin.
- The plasma transports nutrients, hormones, carbon dioxide, salts and urea.
- The white cells combat infections. One type, called phagocytes, engulfs and destroys bacteria.
- The platelets help the blood to clot.
- The heart pumps blood around the body, due to the contraction of its muscular walls.
- The coronary artery supplies the muscle of the heart with blood.
- The mammalian circulatory system is a double circulation, in which blood travels through the heart twice on each circuit of the body.
- The blood leaves the heart in arteries, flows through capillaries in the organs, then back to the heart in veins.
- Blood travels through the heart by entering the atria, passing through to the ventricles, and then being pumped out into the aorta or pulmonary artery.
- The heart has two halves. The right half deals with de-oxygenated blood, the left half with oxygenated blood.
- The main blood vessels entering and leaving the heart are the pulmonary artery, the aorta, the pulmonary vein and the vena cava.
- Valves in the heart ensure the one-way flow of blood through the heart, in the right direction.
- Materials enter and leave the blood through the thin-walled capillaries.
- Certain factors related to lifestyle and genetics can affect the risk of cardiovascular disease.

Exam practice

1 The stages of blood flow through the left side of the heart are shown below, but are not in the correct order:

    A   The left atrium contracts.
    B   Blood moves through the aorta.
    C   Blood moves into the left atrium.
    D   The left ventricle contracts.
    E   Blood moves into the left ventricle.
    F   The atrioventricular valve opens.
    G   The semilunar valve opens.

  a) Arrange the steps in the correct order. [3]
  b) Which blood vessel brings blood to the left side of the heart? [1]
  c) The walls of the left and right ventricles are thicker than those of the atria. Suggest a reason for this. [2]
  d) What is the purpose of the valves in the heart? [1]

2 The graphs below show data relating to blood vessels and the flow of blood around the body.

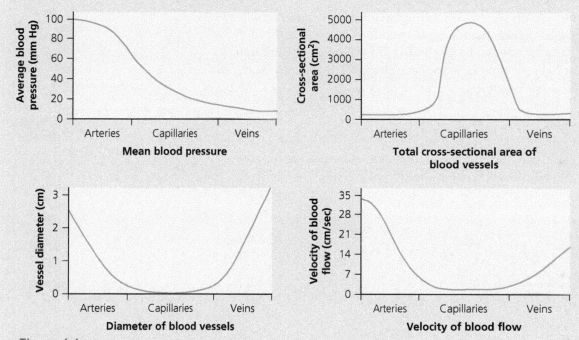

Figure 4.6

  a) Explain why the mean blood pressure falls as the blood travels through the circulatory system. [2]
  b) Explain why the diameter of the capillaries is small, yet their total cross sectional area is very large. [1]
  c) Why is it an advantage to the body that the velocity of blood flow in the capillaries is so low? [1]

Answers and quick quiz 4 online

ONLINE

5 Plants and photosynthesis

Introduction to photosynthesis

Why photosynthesis is important

Photosynthesis is the way green plants and other photosynthetic organisms use chlorophyll to absorb light energy and convert carbon dioxide and water into glucose, producing oxygen as a bi-product. It is important for these reasons:

- Plant life depends on it as a source of food.
- All animals rely on plants for food, either directly or indirectly, and photosynthesis makes that food.
- Photosynthesis produces oxygen which is necessary for respiration. Early plant life added oxygen to the atmosphere, which allowed it to sustain life as we know it.

What plants need to survive

In order to carry out photosynthesis and their other living processes, plants need certain materials from their environment. Figure 5.1 summarises their needs.

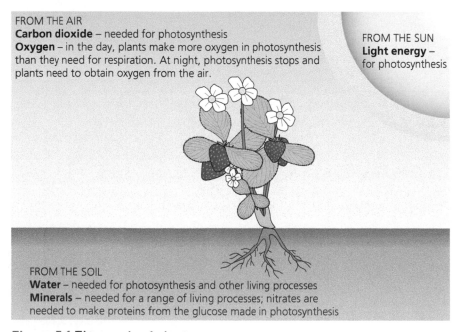

FROM THE AIR
Carbon dioxide – needed for photosynthesis
Oxygen – in the day, plants make more oxygen in photosynthesis than they need for respiration. At night, photosynthesis stops and plants need to obtain oxygen from the air.

FROM THE SUN
Light energy – for photosynthesis

FROM THE SOIL
Water – needed for photosynthesis and other living processes
Minerals – needed for a range of living processes; nitrates are needed to make proteins from the glucose made in photosynthesis

Figure 5.1 The needs of plants.

The process of photosynthesis

Photosynthesis is a complex series of chemical reactions in the chloroplasts of plant cells, but it can be summarised by the following word equation:

carbon dioxide + water → glucose + oxygen

For the process to work, four things are needed:
- **Carbon dioxide** – Glucose is made of carbon, hydrogen and oxygen. The carbon dioxide provides the carbon and oxygen.
- **Water** – This provides the hydrogen needed to make glucose. The oxygen from the water molecules is not needed and is given off as a waste product.
- **Light** – This provides the energy for the chemical reactions in photosynthesis.
- **Chlorophyll** – The green pigment in chloroplasts is chlorophyll, which absorbs the light to provide the energy for photosynthesis.

All the chemical reactions involved in photosynthesis are controlled by **enzymes**, which are available in the chloroplasts of the photosynthesising cells. For photosynthesis to work, the temperature must be suitable for those enzymes to work.

Experimental techniques

There are two types of experiments related to photosynthesis that you need to be familiar with.
- Experiments that show the need for light or carbon dioxide in photosynthesis.
- Experiments that investigate the effect of various factors on the rate of photosynthesis.

The practical techniques for doing these experiments involve the following:

Testing for starch

1 Although plants make glucose in photosynthesis, it is either used or stored as starch very quickly. So to see if photosynthesis has occurred, we test for starch. This is done in the following way:
2 The leaf to be tested is dipped in boiling water to kill the cells and melt the waxy covering (the cuticle) so that liquids can soak into the leaf.
3 The leaf is then placed in boiling alcohol. This removes the green colouring so that the test's colour change can be seen. This is done in a boiling water bath, not by direct heat, because alcohol is flammable (see Figure 5.2).
4 The leaf is dipped in water briefly (the alcohol makes the leaf brittle and the water softens it).
5 Brown **iodine solution** is dripped onto the leaf. A blue–black colour indicates starch.

De-starching leaves

When doing experiments about the formation of starch in photosynthesis, it is important that we know that the starch was formed during the experiment, not before. To ensure this the plant is **de-starched** before the experiment by putting it in a dark cupboard for 24 hours. Any stored starch will then be used up to feed the plant as it will not be able to photosynthesise.

Testing the need for light

The need for light can be tested by covering up part of a leaf with foil or black paper and then testing the leaf for starch after about 24 hours. The part which was covered will not contain starch, but the rest of the leaf will. This is better than comparing different leaves (one in the light, one in the dark) because it is a better fair test. Using one leaf means that all the factors connected with the light and dark areas will be the same.

> **Exam tip**
>
> Remember to state the original colour as well as the colour it changes to when describing a colour change as in the starch test.

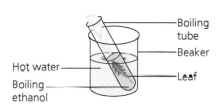

Hot water
Boiling ethanol
Boiling tube
Beaker
Leaf

Figure 5.2 Removing chlorophyll from a leaf.

Removing carbon dioxide and control experiments

If we want to show that carbon dioxide is needed for photosynthesis, we need to be able to remove it from around a leaf and compare starch production with a leaf that has access to carbon dioxide. Such an experiment is shown in Figure 5.3.

Carbon dioxide is removed from leaf B by adding **sodium hydroxide**, which absorbs it, to the flask. Leaf A is a **control experiment**. If, as expected, there is no starch formed in leaf B, we need to know that it was because carbon dioxide was absent and for no other reason. The control experiment is identical to the experimental flask, except that the sodium hydroxide is replaced by an equal volume of water, which does not absorb carbon dioxide. If leaf A contains starch and leaf B does not, we can be certain that it is due to the absence of carbon dioxide around leaf B.

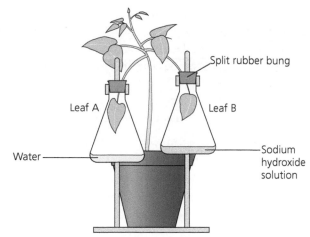

Figure 5.3 Apparatus to investigate if carbon dioxide is needed for photosynthesis.

Using sensors and data loggers

Testing with iodine simply shows that photosynthesis has taken place; it does not measure the rate of photosynthesis. One way of doing this is to monitor either the decrease of carbon dioxide around a leaf, or the increase in oxygen. There are **electronic sensors** which can detect and measure oxygen and carbon dioxide. The sensor needs to be connected to a **data logger**, which records and stores information about the level of the gas over time.

Now test yourself
TESTED

1 Which two essential materials does photosynthesis provide for animals?
2 What is the role of light in photosynthesis?
3 Why do scientists keep the test plant in the dark for 24 hours before starting some experiments on photosynthesis?
4 When testing leaves for starch, why is the leaf boiled in alcohol?
5 Which substance is used to absorb carbon dioxide in some experiments on photosynthesis?

Answers online

More about photosynthesis

Factors affecting the rate of photosynthesis
REVISED

The more photosynthesis a plant can carry out, the more food it will produce and the more it will grow. The following factors affect the rate of photosynthesis.

● **Light intensity**. Light provides the energy for photosynthesis. The higher the light intensity, the higher the rate of photosynthesis (up to a maximum, because eventually the chloroplasts will be absorbing all the light they can).

- **Level of carbon dioxide**. Carbon dioxide provides an essential raw material for photosynthesis, and increasing the level of carbon dioxide will increase the rate of photosynthesis (again, up to a maximum).
- **Temperature.** Photosynthesis is a series of chemical reactions controlled by enzymes. Increasing the temperature will speed up these reactions but if the temperature gets too high (above about 40 °C) the enzymes will denature and the plant will die.

Water is needed for photosynthesis but it hardly ever affects the rate, because if the plant has enough water to stay alive, it has enough for photosynthesis.

> **Exam tip**
>
> When talking about factors that affect the rate of photosynthesis, always refer to light intensity, not just light.

Limiting factors

REVISED

In any set of circumstances, one factor is more important than the others in setting the rate of photosynthesis. This factor is known as the **limiting factor**. In different conditions, any of the factors listed above – light, carbon dioxide or temperature – can be the limiting factor. You can tell if a factor is limiting by increasing it. If the rate of photosynthesis also increases, then the factor was limiting.
- **Light** is limiting at night and at dawn and dusk. In the daytime, even on a very cloudy day, light is not likely to be limiting except in very shaded places.
- **Temperature** can be a limiting factor on cold winter days.
- **Carbon dioxide** will be limiting whenever the light and temperature are not. The level of carbon dioxide in the atmosphere is very low (0.04%).

The use of glucose by the plant

REVISED

Once glucose has been made by photosynthesis, the plant does a variety of things with it:
- It may be **used** directly in respiration to provide energy for the plant.
- It may be **transported** to other parts of the plant (particularly the growing points in the stem and the roots). To do this, it is changed into **sucrose**.
- It may be **stored** by conversion to **starch** or **oils**.
- It can be **transformed** into **cellulose** for cell walls, or into **proteins** for growth. To make proteins, the plant will need a supply of nitrogen from nitrates in the soil.

Summary

- Photosynthesis is the process whereby green plants and other photosynthetic organisms use chlorophyll to absorb light energy and convert carbon dioxide and water into glucose, producing oxygen as a by-product.
- The chemical reactions of photosynthesis are controlled by enzymes.
- Photosynthesis requires carbon dioxide, water and light, together with chlorophyll in chloroplasts to absorb the light.
- The rate of photosynthesis is affected by temperature, levels of carbon dioxide and light intensity.
- A limiting factor is one that is limiting the rate of photosynthesis at a given time.
- Temperature, levels of carbon dioxide and light intensity can act as limiting factors for photosynthesis.
- The occurrence of photosynthesis can be detected by testing a leaf using iodine in potassium iodide solution, which turns blue-black in the presence of starch.
- Glucose produced in photosynthesis can be respired to release energy, converted to starch for storage or used to make cellulose, proteins and oils.

Exam practice

1 A plant was kept in the dark for 48 hours, and then discs were cut from its leaves. These were floated on solutions, some in the light and some in the dark. The leaf discs were then tested for starch. The results are sown in the table below.

Experiment	Solution	Conditions	Result
A	Water	Light	Starch present
B	Glucose	Light	Starch present
C	Water	Dark	Starch absent
D	Glucose	Dark	Starch present

   a) Describe the process by which the leaf discs were tested for starch. [4]
   b) Describe the appearance of a positive result with this test. [1]
   c) Why were the leaves kept in the dark for 48 hours before the start of the experiment? [1]
   d) Explain the result of experiment A. [3]
   e) Explain the result of experiment D. [2]

2 A hydrogen carbonate indicator can be used to detect levels of carbon dioxide in solution. Carbon dioxide is an acid gas, and acids turn the orange-red bicarbonate indicator yellow. If carbon dioxide is reduced, the solution becomes alkaline and turns purple. Five test tubes were set up as shown in the diagram, and left for 3 hours.

The colours of the indicator in the five tubes after 3 hours were as follows:

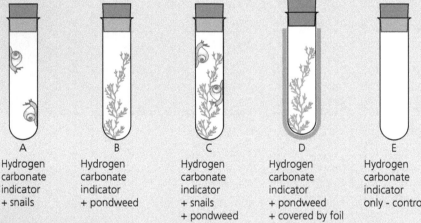

A	B	C	D	E
Hydrogen carbonate indicator + snails	Hydrogen carbonate indicator + pondweed	Hydrogen carbonate indicator + snails + pondweed	Hydrogen carbonate indicator + pondweed + covered by foil	Hydrogen carbonate indicator only - control

Test tube	Colour of indicator
A	Yellow
B	Purple
C	Orange-red
D	Yellow
E	Orange-red

   a) What process is responsible for the colour change in test tube A? [1]
   b) What process is responsible for the colour change in test tube B? [1]
   c) Explain the results in test tubes C and D. [4]
   d) What is the purpose of the control test tube E? [1]

Answers and quick quiz 5 online

ONLINE

6 Ecosystems, nutrient cycles and human impact on the environment

Energy flow in the environment

Energy in food chains

REVISED

Sunlight is the source of all the energy on the planet. This energy passes from organism to organism by means of **food chains**. Plants are the first links in all food chains because they are **producers** – they change energy in sunlight into stored chemical energy, although plants only manage to capture about 5% of the energy in sunlight. When plants are eaten by **herbivores**, some of the energy is passed to the next link, the **consumers** in the food chain. When the herbivore is eaten by a **carnivore**, the process of energy transfer is repeated. Energy passes in this way from carnivores to scavengers and **decomposers**, which feed on dead organisms.

The energy transfer at each stage is not 100% efficient. Energy is lost in the following ways:
- A lot of light energy misses plants completely.
- Some of the light which does hit a plant is reflected or goes through the leaves and so is not absorbed.
- Each stage in a food chain uses some of the energy it takes in for its own purposes (e.g. movement, cell repair, reproduction). That energy is dispersed and cannot be passed on.
- Organisms rarely eat every bit of another organism (e.g. a herbivore will often not eat the roots of a plant) and so do not take in all the energy available.
- Even what is eaten is often not completely digested, so some energy never gets into the feeding organism.

H As energy is lost at each stage of a food chain, the numbers of organisms that can exist at each level of a food chain goes down as the food chain progresses.

> **Exam tip**
>
> The source of *food* in a food chain is green plants (producers), but the source of *energy* is sunlight. If asked about food chain sources, read the question very carefully and answer appropriately.

Trophic levels

The stages in a food chain are called trophic levels, and each level is given a specific name. The following is an example of a food chain.

Plant plankton → animal plankton → small fish → larger fish → tuna → human

The names of the trophic levels in this food chain are given in Table 6.1.

Table 6.1 Stages in a food chain.

Organism	Name of trophic level
Plant plankton	Producer
Animal plankton	First stage consumer
Small fish	Second stage consumer
Large fish	Third stage consumer
Tuna	Fourth stage consumer
Human	Fifth stage consumer

Note that this is a very long food chain. Most food chains do not go further than the third or fourth stage consumer.

Food webs

In nature, food chains often interlink, because most organisms eat a lot of different things and are eaten by many different animals, too. Interlinked food chains are called **food webs**. Figure 6.1 shows an example, but even this food web is an oversimplification of all the feeding relationships that would exist in this environment.

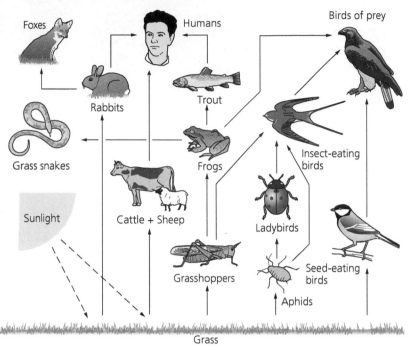

Figure 6.1 **An example of a food web.**

Ecological pyramids

Feeding relationships can be illustrated as pyramids (Figure 6.2 and Figure 6.3). The width of each block in the pyramid is an indication of the number (or mass) of that type of organism at that feeding level.

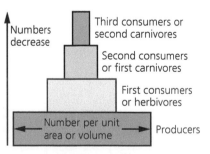

Figure 6.2 **A pyramid of numbers for a grassland food chain.**

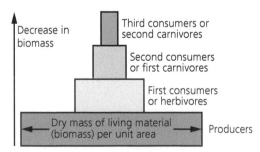

Figure 6.3 **A pyramid of biomass for the same grassland food chain.**

There are two types of pyramid:
1 A **pyramid of numbers** shows the number of organisms per unit area or volume at each feeding level.
2 A **pyramid of biomass** shows the **dry mass** of organic material per unit area or volume at each feeding level.

Pyramids of biomass give a more accurate picture than pyramids of numbers. Pyramids of numbers are sometimes not pyramid shaped. Sometimes the producers (e.g. trees) are much larger than the herbivores that feed on them (e.g. insects). One tree can support thousands of insects, so in a pyramid of numbers the bottom block, representing producers, is narrower than the block for the first stage consumers (Figure 6.4). A tree weighs a lot more than all the insects feeding on it put together, so a pyramid of biomass will be pyramid shaped as expected. A similar situation can arise when an animal is fed on by many smaller animals (e.g. a dog with fleas).

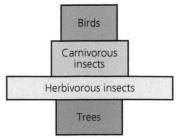

Figure 6.4 Example of a pyramid of numbers that is the 'wrong' shape.

> **Dry mass:** The mass of a material after all the water has been removed.

Conservation

REVISED

Human needs sometimes conflict with conservation of the environment. Humans sometimes have to destroy habitats in order to meet essential needs, for example:

- People need places to live and work, but building houses, roads etc. destroys the habitat.
- Farmland completely changes the natural habitat, but people need food.
- To meet energy needs, people have to build power stations and tidal barrages.
- To provide essential drinking water, rivers may be dammed and areas deliberately flooded.
- The issues vary according to the situation. How intense the human need is will vary. The need to conserve the natural environment will be stronger in certain situations, e.g.
  - The habitat may contain rare or endangered species.
  - The habitat may have a very large variety of species in it, so that many populations will be affected if it is destroyed.
  - The habitat itself may be unusual or rare, making its conservation more important.

Farming practices

REVISED

Intensive farming is an agricultural system that aims to produce a maximum yield from the land available. It applies to both animals and plants, and it involves the use of chemicals such as pesticides and fertilisers to increase yield and to control diseases. Farm animals may be kept indoors in restricted conditions so that as many as possible can be kept in a given space. This is called battery farming. There are advantages and disadvantages of intensive farming, but people have to decide their own opinions about it. Scientific data cannot decide an ethical issue, but can provide the accurate information people need to help them make their decisions.

The advantages of intensive farming are:
- The yield is high because more livestock animals or crop plants can be kept and conditions can be controlled. Therefore, the food is cheaper to produce and more profitable for the farmer, which may keep him or her in business. If farms go out of business, the UK becomes less self-sufficient in food.
- Food is cheaper in the shops, allowing people to choose a healthier diet, even on a tight budget.
- By increasing the yield, it allows the UK to grow more food, to meet the needs of a growing population.

The disadvantages are:
- The chemicals used (for example, pesticides and antibiotics used to control disease in livestock) could enter the human food chain and get into our bodies.
- The chemicals can cause pollution and harm wildlife other than pests.
- Natural environments are destroyed. For example, hedgerows are uprooted to make large fields suitable for intensive farming.
- Although no-one can really know what an animal 'feels', it is likely that intensive farming causes animals stress and discomfort. Their quality of life is poor.

> **Exam tip**
>
> This section involves ethical issues. You cannot really learn it, because exam questions will normally present a scenario which you may not have come across. You need instead to make sure you understand the sorts of issues involved, so that you can apply your understanding to any context.

Now test yourself

TESTED

6 Which biological process removes carbon dioxide from the air?
7 In what form do plants absorb nitrogen?
8 What do nitrogen-fixing bacteria do?
9 Which enzyme in bacteria breaks down urea into ammonia?
10 State one advantage of intensive ('battery') farming of chickens.

Answers online

Pollution

Types of pollutant

REVISED

A **pollutant** is something that has been added to the environment and which damages it in some way. Some common pollutants are:
- solid or liquid chemicals, such as oil, detergents, fertilisers, pesticides and heavy metals
- gaseous chemicals, such as carbon dioxide, methane, chlorofluorocarbons (CFCs), sulfur dioxide and nitrous oxides
- human and animal sewage
- noise
- heat
- non-recyclable household waste.

Detecting pollution

Water pollution can be detected by a fall in the **oxygen level** or a change in **pH** in streams and rivers. Scientists can often judge the level of pollution by using **indicator species** – plants and animals which have a known tolerance of pollution. Some will always be absent once pollution reaches a certain level, while others will only be found when pollution exceeds a certain level.

There are many different indicator species which are found in fresh water and can be used to test pollution. **Lichens** are a good indicator of polluted air, different species being intolerant or very tolerant of air pollution.

Pollutants in food chains

Some harmful chemicals cannot be broken down in the body. If they are eaten, they remain in the tissues. Any animal that eats an organism that contains such a harmful chemical will then take it in. Organisms eat large numbers of prey from the trophic level below them, and so they will take in a lot of the chemical. This happens at every stage in the food chain and so the higher up the food chain we go, the greater the level of the harmful chemical in the body. The level can get so high that it harms or kills the organism. Figure 6.5 shows an example of how a harmful chemical (the insecticide DDT) builds up in a food chain (you do not need to learn this example – it is for illustration only).

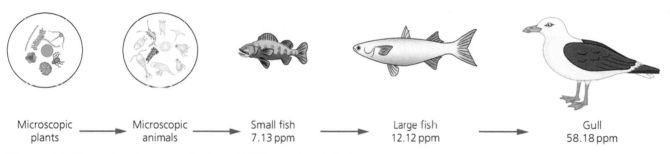

| Microscopic plants | → | Microscopic animals | → | Small fish 7.13 ppm | → | Large fish 12.12 ppm | → | Gull 58.18 ppm |

Figure 6.5 Example of accumulation of DDT in a food chain.

The organisms which are at the top of any food chain are most likely to be harmed. As humans are at the top of many food chains, we are particularly at risk.

Eutrophication

Sewage and fertilisers sometimes get into streams and rivers from farmland, washed from the soil by rain. This can cause **eutrophication**, which happens in the following way:

1 The sewage or fertiliser causes an increase in the growth of microscopic plants.
2 These plants have short lives and soon die.
3 Bacteria feed on the dead plants in large numbers.
4 The oxygen level in the water drops because of bacterial respiration.
5 Animals, such as fish, which need a lot of oxygen, die due to the low oxygen levels.
6 In small bodies of water, such as ponds, the microscopic plants may form a complete blanket over the surface, so cutting off the light that the plants at the bottom of the pond need to survive.

Now test yourself

11 Define the term 'pollutant'.
12 What can scientists measure to detect the level of pollution in river water?
13 What name is given to a species which can only live within a narrow range of pollution?
14 A lake is polluted by a harmful chemical. Why is this more likely to damage the population of birds of prey in the area, rather than the small fish?
15 Which organisms remove oxygen from the water when eutrophication occurs?

Answers online

Summary

- Food chains and food webs show the transfer of energy between organisms and involve producers, first stage consumers (herbivores), second and third stage consumers (carnivores), and decomposers.
- Some food chains involve fourth or even fifth stage consumers, but this is unusual.
- At each stage in the food chain energy is used in repair and in the maintenance and growth of cells, while energy is lost in waste materials and respiration.
- Pyramids of numbers and biomass indicate the numbers or mass of the organisms at different trophic levels.
- The efficiency of energy transfers between trophic levels affects the number of organisms at each trophic level.
- Human requirements for food and economic development need to be balanced with the needs of wildlife.
- Intensive farming methods – such as using fertilisers, pesticides, disease control and battery methods to increase yields – have both advantages and disadvantages.
- Indicator species and changes in pH and oxygen levels may be used as signs of pollution in a stream and lichens can be used as indicators of air pollution.
- Harmful chemicals that enter the food chain accumulate in animal bodies and may reach a toxic level.
- Untreated sewage and fertilisers may run into water and cause rapid growth of plants and algae, which then die and are decomposed. The microbes that break them down increase in number and their respiration can use up the oxygen in the water, causing harm to animals living there. This is known as eutrophication.

Exam practice

1 Animal waste was accidentally discharged from a farm into a nearby river. The effect of this discharge on various factors in the river is shown in the graph below.

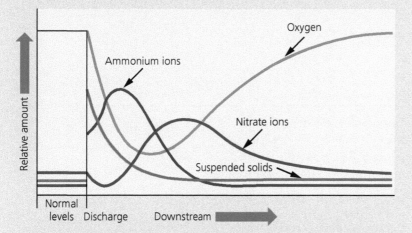

a) Explain the following events seen in the graph.
   i) The rise in ammonium ions at the point of the discharge. [1]
   ii) The decrease in oxygen levels downstream from the discharge. [3]
   iii) The rise in nitrates shortly after the discharge. [2]
   iv) The drop in the level of suspended solids. [1]
b) If the discharge had been into still water rather than a river, the decrease in oxygen would have been greater and would have lasted for longer. Suggest a reason why the problem is less severe in moving water. [1]
c) What name is given to the process which results in oxygen levels falling after pollution by fertiliser? [1]

2 Below is an example of a food chain.

   Grass → mouse → snake → eagle
a) What is the source of energy in all food chains? [1]
b) In the example, the grass is referred to as a producer. What trophic level does the snake belong to in this food chain? [1]
c) Suggest why all the energy in the grass is not available to the mouse? [2]
d) The land on which the grass grows becomes polluted with a heavy metal. Which organism in the food chain will be most at risk from poisoning by the heavy metal? Give a reason for your answer. [2]
e) Below are some pyramids of number. Which one is the most accurate for the example food chain given? [1]

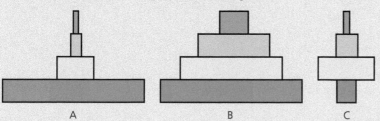

A   B   C

Answers and quick quiz 6 online

ONLINE

7 The variety of life

Classification and biodiversity

The ways living things vary

Living organisms vary in size and complexity, and have different structural and functional features. Some are just one cell big, while the most complex consist of trillions of cells, organised into tissues, organs and organ systems. The most familiar living things fall into two groups:

- Plants – flowering and non-flowering.
- Animals – vertebrates (which have a backbone) and invertebrates (which do not).

There are other organisms, principally bacteria, which do not belong to either of the above groups.

Classifying living organisms

Scientists classify living things into groups which have similar features and characteristics. They do this for the following reasons:

- It makes it easier to identify organisms – if you can identify the group an organism belongs to, it narrows down the search to identify it. The smaller the group, the easier it becomes.
- It helps scientists communicate with each other. For example, if you describe an animal as an insect, everyone knows what type of animal you are talking about, and that it will have features associated with that group.

To help scientific communication further, organisms are given scientific names, which are agreed internationally. This avoids any confusion that might arise if names in the local language were used. The scientific name is always two words, e.g. *Homo sapiens* (human), *Ranunculus acris* (meadow buttercup).

Adaptation to the environment

One reason even closely related species can show quite a lot of differences is that, through natural selection, species become adapted to their environment. Features develop that help the organisms to survive, so if similar organisms live in different environments, they will adapt in different ways. These adaptations are of two sorts:

Morphological adaptations are structural adaptations of the organism (either internal or external) – for example, colour of fur, leg length, petal shape, reduced size of appendix and so on.

Behavioural adaptations could include the time of day when an animal is active or the type of food it eats. Plants have very limited 'behaviour', so this mostly applies to animals.

There are some common types of adaptation that animals and plants show, listed below:

- Adaptations for camouflage. Colouring to match their environment.
- Adaptation to climate. Thick or thin fur in animals to retain heat or lose it; layers of fatty insulation; large ears to lose heat to the environment or small ones to avoid it. Plants often have adaptations (e.g. thick waxy cuticle on their leaves) to avoid water loss in dry environments, or deep roots to reach water.
- Adaptations for catching prey or avoiding being caught. Besides camouflage, animals may have long legs so that they have a longer stride and can run faster.

Figure 7.1 shows an example of adaptations in an animal.

Exam tip

When asked about adaptations, you will probably be given an animal or plant and details of their environment. There is no point learning the specific adaptations of any one animal or plant, you need to understand the principles and apply them.

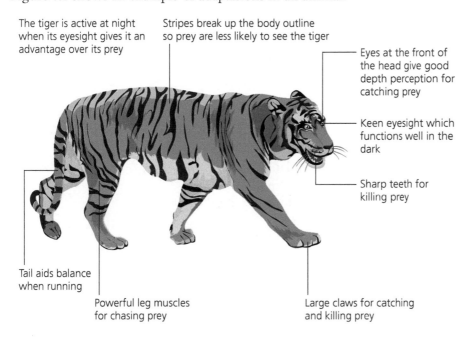

The tiger is active at night when its eyesight gives it an advantage over its prey

Stripes break up the body outline so prey are less likely to see the tiger

Eyes at the front of the head give good depth perception for catching prey

Keen eyesight which functions well in the dark

Sharp teeth for killing prey

Tail aids balance when running

Powerful leg muscles for chasing prey

Large claws for catching and killing prey

Figure 7.1 The tiger's adaptations not only suit it to its way of life as a predator, but also to the environment (it is successful in the jungle, but its stripes would provide much less effective camouflage in grassland).

The needs of living organisms

REVISED

An environment must have certain factors available if it is to support life. These are shown in Table 7.1.

Table 7.1 Resources needed by living things.

Resource	Needed by
Light	Plants to make food for energy
Food	Animals for energy
Water	All living organisms, for the chemical reactions that take place in cells
Oxygen	All living organisms that respire aerobically, to break down food and release its energy
Carbon dioxide	Plants for photosynthesis
Minerals	All living organisms – specific minerals are needed for particular chemical reactions that take place in cells

Competition for resources

The organisms in any habitat compete for essential resources and those that are more successful at this competition will survive better than the rest.

- **Competition** always takes place between members of the same species, because they all require the same things (for example, they eat the same food), but it also occurs between organisms of different species with similar needs.
- Competition puts a limit on the potential size of a population, although other factors are also important – for example, **predation**, **disease** and **pollution**.
- Competition only occurs when a particular factor (e.g. food) is in limited supply. If any essential resource is in unlimited supply (which rarely occurs) there will be no competition.
- When there is competition, the population cannot reach a size that it might have done if the resources were infinite.
- Predation, disease and pollution all contribute to the death rate in a population, and so also limit its size.

Biodiversity

Biodiversity is the number of different species (of all types) in a particular area. It also relates to the number of animals and plants of each species that are there. The 'area' concerned is of no fixed size – you could talk about the biodiversity on a sea shore, or in Wales, or in Europe, and so on. Biodiversity is a good thing, for a number of reasons.

- It leads to stable environments that can resist potentially harmful situations (see p. 45).
- Habitats which have a great variety and number of animals and plants are also more interesting to humans for both scientific and leisure purposes.
- Many important medicines have been developed from chemicals in plants, and many more are probably undiscovered. The more different types of plants there are available, the more potential there is for extracting new drugs.
- Hundreds of years of selective breeding in domestic and farm animals and crops have sometimes resulted in the loss of resistance to certain diseases. If ancient breeds have been conserved, they can strengthen current breeds, or re-introduce disease resistance, by crossbreeding them with the modern variety.

Now test yourself

1 Why do scientists give each species a scientific name?
2 What is a morphological adaptation?
3 The hair of the arctic hare changes colour from brown to white when winter comes. Suggest how this seasonal colour change is an adaptation to the environment.
4 In sand dunes, there are relatively few plants, all of which are short. Suggest why competition for light is unlikely to be a factor in sand dunes.
5 Name four factors that might limit the size of a population.

Answers online

Biodiversity and stability

The food web (Figure 7.2) is from an environment that shows some biodiversity. Several species are present, and the numbers of each are quite high.

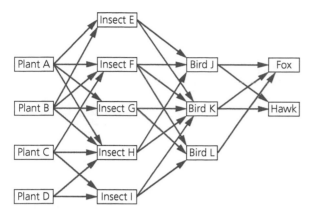

Figure 7.2 A food web.

Suppose a change in the environment leads to a large drop in the population of insect G:

- Birds K and L would have less of insect G to feed on, but each has other insects they can eat, and so the effect on them will be quite small.
- Even though more of the other insects might be eaten by birds, there will be less competition for food without insect G, which would balance out that effect.

The more biodiversity there is in an area, the more options each species has if a source of food is wiped out, and so the environment is more stable.

Maintaining biodiversity

There are various ways in which endangered species can be conserved, or biodiversity can be maintained, locally or nationally:

- breeding and release programmes to boost populations (e.g. in zoos)
- active conservation of habitats of threatened species
- re-creation of habitats that have declined (planting, landscaping, and so on)
- control of invasive species that may be spreading and pushing out other species
- legislation to protect habitats or individual species
- controlling pollution or other factors that might be threatening species or their habitats.

Legislating (that is, making laws) to protect habitats or species can be difficult in some cases, as wildlife needs may conflict with human needs. In addition, the living world is very varied and it can be quite difficult to frame legislation that will cover all situations fairly.

Getting data about biodiversity – sampling

Unless the environment studied is very small, the whole of it cannot be explored to find all the animals and plants living there. The only way we can get an idea of numbers is to take a **sample** of the environment. A small area is surveyed, and used to estimate the numbers in the whole environment.

There are three principles related to how to choose a sample:

- The sample must be **big enough** to be representative (i.e. typical) of the area as a whole.

> **Exam tip**
>
> The principles of sampling described here apply not only to measuring biodiversity. They also apply to surveys done on human populations, although some surveys are targeted at specific groups of the population and so are not entirely random. Exam questions could be asked about sampling in the context of surveys.

- The sample must be **random**, so that it is free from any unconscious bias on the part of the data collector.
- The method of sampling must not affect the results (e.g. the presence of a human may frighten some animals away from the sampling area).

Getting data about biodiversity – quadrats

A quadrat is a wooden or metal square frame with sides of a standard length. It can be used to study the biodiversity of an area. It is used in the following way:

1 The area to be sampled is divided up into a numbered grid. Each square of the grid is the size of one quadrat.
2 A random number generator is used to choose the numbers of the squares to be sampled. This ensures that the sample is random. The scientist decides how many squares to sample, so that the sample is typical of the whole area. The more varied the area, the greater the number of squares sampled.
3 The quadrat is placed on each of the numbered squares in turn.
4 The species within the quadrat are identified, counted and recorded.
5 When the whole sample is finished, the numbers are used to calculate the numbers of each species in the area.

Estimating biodiversity – a worked example

An area of 2,000 m^2 is sampled, counting the number of dandelion plants. The sample is 100 quadrats, each of which has a side of 0.5 m. The total count of dandelions is 70. Estimate the number of dandelion plants in the area.

1 First, work out the area that has been sampled.
Each quadrat has an area of 0.5×0.5 m^2 = 0.25 m^2
There are 100 quadrats, so the total area sampled = 100×0.25 m^2 = 25 m^2
2 Now calculate the total number of dandelions as follows:
Number of dandelions = Number of dandelions counted × Total area/ Area sampled
Number of dandelions = $70 \times \dfrac{2000}{25} = 5600$

Getting data about biodiversity – animal populations

Quadrats cannot be used to sample animal populations where the animal moves. There would be a possibility of counting the same animal more than once in different sample areas, or of missing animals which have temporarily moved out of the sample area.

H To determine the size of an animal population, the **capture–recapture technique** can be used. The technique works like this:

- A number of individuals of a particular species are captured.
- These animals are marked in some way so they can be distinguished from the rest of the population. They are released back into the habitat.
- Later, another sample of the species is captured. The proportion of marked individuals in the second sample will be the same as the proportion of the initially marked individuals in the total population.

H The total population can be estimated using the equation:

$$N = MC/R$$

where:

N = estimate of total population size

M = total number of animals captured and marked on the first visit

C = total number of animals captured on the second visit

R = number of animals captured on the first visit that were then recaptured on the second visit.

Animal populations, alien species and biological control

H Capture–recapture technique calculations

The capture–recapture technique will only give accurate results when the following conditions apply:
- Sufficient time must have elapsed between the taking of the two samples for the marked individuals to mix with the rest of the population.
- There must be no large-scale movement of animals into or out of the area in the time between the two samples.
- The marking technique must not affect the survival chances of the animal (for example, making it easier for a predator to see it).
- The marking technique must not affect the chances of recapture by making the marked individuals more 'noticeable' to the collector.

Capture–recapture technique – a worked example

A scientist wanted to estimate the number of snails in a garden. She collected 50 snails and marked them with white paint underneath the shell. She then released them. 2 days later, she collected another sample of 50 snails from the garden. Twenty of them were snails she had already marked:

$$N = MC/R$$

N = estimate of total population size

M = total number of animals captured and marked on the first visit = 50

C = total number of animals captured on the second visit = 50

R = number of animals captured on the first visit that were then recaptured on the second visit = 20

$N = 50 \times \dfrac{50}{20} = 125$ snails in the garden

Alien species and biological control

An alien species is one that is not normally found in a country or geographical area, but has been introduced from a different part of the world. Introducing an alien species can create problems, some of which are listed below:

- The alien species may have no predators in the area, and its population may grow out of control.
- It may compete with an existing species, causing that species to die out in the area.
- It could prey on existing species, reducing their number.
- It might carry a disease to which it has immunity, but existing populations do not.

Some species are deliberately brought in to control pest species. This is an example of **biological control**. Biological control involves using living organisms (often predators) instead of chemical pesticides. In the early days of biological control, the process sometimes went wrong, and the introduced predator itself caused a problem, e.g. some populations grew out of control because they had no predators, or ate species apart from the pest species.

Scientists now understand the possible problems of introducing biological control agents and detailed research and extensive trials are now used before introducing any control species.

Now test yourself

TESTED

6 What role can zoos have in maintaining the biodiversity of the planet?
7 When using quadrats to measure biodiversity, why must sampling be random?
8 Why would quadrats be an inappropriate technique for sampling the butterfly population of an area?
9 Students sampled the beetle population of an area using the mark and recapture technique. They marked the beetles they caught with luminous paint on their backs. Why might this be an inappropriate marking technique?
10 What is 'biological control'?

Answers online

Summary

- Living organisms show a range of sizes, features and complexities.
- Plants are broadly divided into flowering and non-flowering groups; animals are broadly divided into vertebrates and invertebrates.
- A scientific system for identification is needed to simplify and better organise the study of living things.
- Organisms with similar features and characteristics are classified into groups.
- Organisms are given scientific names to avoid confusion that could be caused by local or national names.
- Organisms have morphological and behavioural adaptations that help them to survive in their environment.
- Individual organisms need resources from their environment – for example, food, water, light and minerals.
- The size of a population may be affected by competition for resources, as well as by predation, disease and pollution.

- Biodiversity is the variety of different species in an area, as well as the number of individuals of those species.
- Biodiversity is important in maintaining the stability of an ecosystem.
- Biodiversity (and endangered species) can be protected by local and national measures.
- Quadrats can be used to investigate the abundance of species.
- Samples must be randomly placed, and sufficiently large, to be an accurate representation of the total area being investigated.
- The capture–recapture technique can be used to estimate population size for mobile organisms.
- The method of marking must follow certain criteria so as not to bias the results.
- The introduction of alien species can have harmful effects on local wildlife.
- Biological control agents can be effective, but there are issues surrounding their use that must be considered.

Exam practice

1 Customs regulations make it illegal to bring any living plant or animal into the country without special permission. Explain why the introduction of a species from another country could damage British ecosystems. [6]

2 Scientists grew two species of clover (*Trifolium fragiferum* and *Trifolium repens*) separately and together. Both species have similar requirements. *T. fragiferum* grows to be taller than *T. repens*, but more slowly. The graphs show how well they grew in each situation. Leaf area index is a measure of the growth of the population.

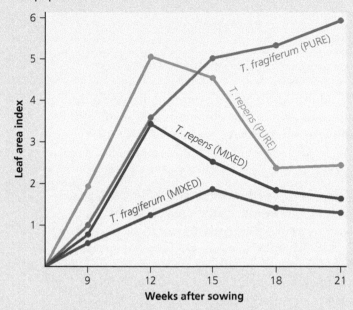

a) *Trifolium fragiferum* is also known as the strawberry clover, and *Trifolium repens* as the white clover. Suggest why the scientists used the scientific names rather than their common names. [1]

b) Describe the pattern of growth of *T. repens* when grown on its own over the 21-week period. [3]

c) Describe how the growth of *T. fragiferum* changed when it was grown mixed with *T. repens*, compared with when it was grown on its own. [2]

d) Explain how the graphs indicate that the two species of clover compete with one another. [1]

e) Using the information in the question, suggest why neither one of the clover species completely eliminates the other one in competition. [1]

3 Scientists were investigating the number of woodlice in a small wood using the capture-recapture technique. They went into the wood and captured 200 woodlice. They marked them underneath their bodies with white paint and then released them. 5 days later, they went back to the wood and captured another 200 woodlice. In this sample, 5 had white paint on them.

The number of woodlice in the woodland can be calculated using the following formula:

$N = MC/R$

N = estimate of total population size
M = total number of animals captured and marked on the first visit
C = total number of animals captured on the second visit
R = number of animals captured on the first visit that were then recaptured on the second visit

a) Calculate the estimated number of woodlice in the wood. [2]

b) Why did the scientists wait 5 days before taking a second sample? [1]

c) Why did they mark the woodlice underneath their bodies, rather than on top? [2]

d) Suggest why a sample of 50 woodlice would have been too few in this location. [2]

Answers and quick quiz 7 online

ONLINE

8 Cell division and stem cells

Genes, chromosomes, cell division and stem cells

Chromosomes and genes

Chromosomes are found in the nucleus of every cell, but are only visible when a cell is about to divide. The chromosome is made of the chemical **deoxyribonucleic acid** (**DNA**). Sections of DNA that control features of the organism are called **genes**. A chromosome contains many genes.

The number of chromosomes in a cell varies between different species – human body cells have 46, 23 from each parent, so the 46 chromosomes consist of 23 pairs of chromosomes. The pairs are not identical, although they look the same and they have the same genes. They are not identical because the form of the gene (called the **allele**) can vary. The full set of human chromosomes, arranged in their pairs, is shown in Figure 8.1. These are the chromosomes of a male, and males have one 'pair' of chromosomes that do not look the same (the X and Y chromosomes). Females have two X chromosomes.

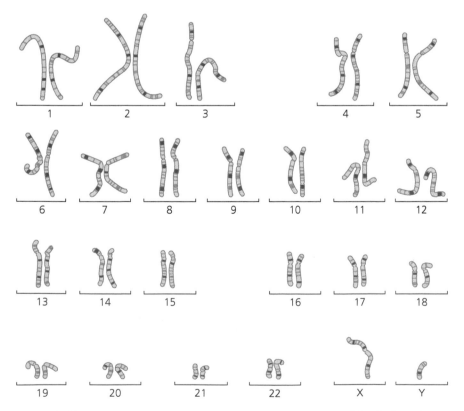

Figure 8.1 The chromosomes of a human male, arranged in their pairs.

Cell division

In multicellular organisms, cell division results in growth, and repair and replacement of old or damaged cells and tissues. The type of cell division that occurs in these processes is called **mitosis**, where one cell (the 'mother' cell) divides to form two new ('daughter') cells. The **daughter cells** are genetically identical to the mother cell, because the chromosomes replicate (copy) themselves and one copy is passed into each of the two new cells. The number of chromosomes is the same in the daughter cells as in the mother cell. Mitosis is the normal type of cell division, but there is another type. This is called **meiosis**, and it only occurs when sex cells (**gametes**) are formed.

When forming gametes it is important that the sperm and egg cells do not have 46 chromosomes, otherwise the baby would have 92 chromosomes, and would not develop normally. In meiosis, although the DNA and chromosomes duplicate as in mitosis, four new cells are formed instead of two, and each cell receives a half set of chromosomes. In humans, therefore, the sperm and egg cells each have 23 chromosomes, so that a new zygote will have 46 chromosomes as it should. The chromosomes come in pairs that are not identical, because during the process of meiosis the gametes get one chromosome from each pair. The new cells in meiosis, unlike mitosis, are therefore not genetically identical.

The way cells divide in mitosis and meiosis is shown in Figure 8.3, and the differences between mitosis and meiosis are shown in Table 8.1.

Mitosis is normally carefully regulated to allow cells time to grow and become specialised. Sometimes this control fails, mitosis gets out of control and the cells produce cancerous tumours.

Table 8.1 Comparing mitosis and meiosis.

Mitosis	Meiosis
Occurs in all body cells except those forming gametes	Occurs only in gamete-forming cells
Daughter cells are genetically identical	Daughter cells are genetically different
Two daughter cells are formed	Four daughter cells are formed
Daughter cells have a full set of chromosomes	Daughter cells have a half set of chromosomes

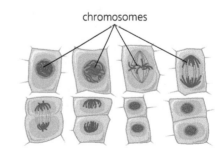

Figure 8.2 Chromosomes in onion root cells.

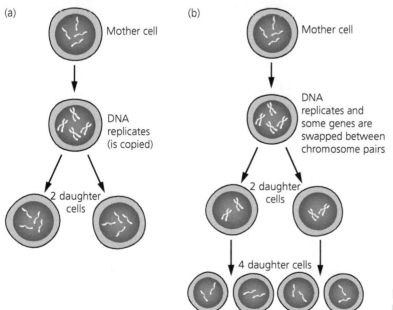

Figure 8.3 The process of (a) mitosis and (b) meiosis.

Stem cells

When a plant or animal embryo is formed and starts to grow, the cells all appear the same. Eventually, the cells start to **differentiate** – to become specialised in some way, as a liver cell, a nerve cell or an epidermal cell, for example. Once they have differentiated, if they then divide, they can only form similar cells to themselves. A liver cell can never become a nerve cell. The undifferentiated cells in the embryo, though – called **stem cells** – can become any cell at all. Scientists can take stem cells from an embryo and grow them into types of cells that can be used to repair or replace damaged tissue. The use of stem cells has proved quite controversial.

There are great potential advantages of using embryonic stem cells, including the following:

- They could be used to treat diseases, such as cancer and type 1 diabetes, which affect or kill millions of people.
- They could be used to repair damaged tissue, such as in brain damage, or spinal cord injuries which can cause paralysis.

People object to the use of stem cells because the embryo which provides the stem cells is destroyed in the process. The issues connected with this argument are:

- Any embryo has the potential to develop into a human being, and so the process destroys life. However, the embryos used are those rejected during fertility treatment, so even if they were not used they would never get the chance to develop.
- Some people are worried that embryos might be created specifically to produce stem cells, so it would be rather like 'farming' human beings (but this does not happen at the moment).

There are possible alternatives to using embryonic stem cells. Stem cells can be found in adults (for example, in the bone marrow inside bones). These cells are in mature tissues but, unusually, have not lost the ability to differentiate into different cells. The problem is that they cannot develop into as many different types of cell as embryonic stem cells can.

Stem cells can also be collected from the blood from the umbilical cord at birth. These would be very useful in treating that baby in later life, as the cells would be genetically identical to its own cells, and so would not be attacked by the immune system.

> **Exam tip**
>
> If you are asked about objections to the use of stem cells, do not just say that 'religious groups object' or that there are 'religious objections.' Always explain what those objections are.

Stem cells in plants

Stem cells also exist in plants. They are found in the growing areas (called **meristems**) in roots and shoots. They can be used to cultivate new plants, but have no medical uses.

Now test yourself

1 Which is bigger, a chromosome or a gene?
2 A liver cell in a hedgehog has 90 chromosomes. How many chromosomes would you find in
   a) a hedgehog kidney cell and
   b) a hedgehog egg cell?
3 When a cell divides by meiosis, how many new cells are formed?
4 What does the term 'differentiate' mean?
5 Why are adult stem cells not as useful as embryonic stem cells for repairing damaged tissue?

Answers online

Mitosis and meiosis

- Cell division by mitosis enables an organism to grow, and to replace and repair cells.
- In mitosis, the number of chromosomes remains constant and the daughter cells are genetically identical to the mother cell.
- Sex cells (gametes) are formed by a different form of cell division called meiosis.
- In meiosis, the number of chromosomes is halved and the daughter cells are not genetically identical.
- Mitosis produces two daughter cells, while meiosis produces four.
- If mitosis is uncontrolled it can result in cancer.
- In mature tissues, the cells have usually lost the ability to differentiate into different forms.
- In both plants and animals, certain cells, called stem cells, are capable of differentiating into different forms of cell.
- Human stem cells have the potential to replace damaged tissue and could be the basis of treatment for a variety of diseases and conditions.
- Human stem cells can be obtained from embryos and from adult tissues.

Exam practice

1 A skin cell from a cow contains 60 chromosomes. It divides to form new skin cells.
   a) After the cell has divided four times, how many chromosomes will be present in each of the new cells? [1]
   b) What chemical are chromosomes made from? [1]
   c) What type of cell division will have produced the new skin cells? [1]
   d) How many chromosomes would you find in a sperm cell of this breed of cattle? [1]

2 The diagram on the right shows a cell from a fruit fly.
   a) What type of cell division has produced this cell? [1]
   b) Explain how you arrived at your answer for (a). [2]
   c) For each of the following statements, state whether it applies to mitosis, meiosis or both.
      i) The cells produced are genetically different [1]
      ii) When division is completed, two new cells have been formed. [1]
      iii) The daughter cells have the same number of chromosomes as the cell which formed them. [1]
      iv) The daughter cells have the same number of chromosomes as each other. [1]

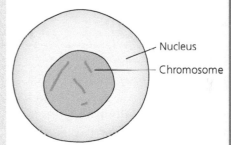

Nucleus

Chromosome

3 Stem cells have many potential uses in medicine, and can be obtained from embryos and from certain adult tissues. Describe what stem cells are and why they could be useful in treating certain conditions. Explain the advantages and disadvantages of using embryonic stem cells compared to adult stem cells. [6]

Answers and quick quiz 8 online

ONLINE

9 DNA and inheritance

DNA and genes

How DNA controls the cell

REVISED

The nucleus controls the activities of the cell, via a chemical called DNA (deoxyribonucleic acid). The way DNA does this is as follows:
- DNA forms a chemical 'code' which acts as instructions for the cell to make proteins.
- The make-up of the DNA molecule varies in different individuals, although it is made up of the same basic components.
- All living processes are a series of chemical reactions in cells.
- All of these chemical reactions are controlled by enzymes.
- All enzymes are proteins.
- The chemical reactions that a cell can carry out are therefore determined by the make-up of the DNA in that individual.

The chemical code in DNA controls which amino acids are joined together in which sequence to make different proteins. In addition to enzymes, other important molecules in the body are proteins, e.g. antibodies and most hormones. DNA can make copies of itself, so that when a cell divides, a copy of the DNA can be put into the nucleus of each of the new cells.

> **Exam tip**
>
> Generally, it is not a good idea to use abbreviations when answering exam questions. However, you do not have to call DNA by its full name (unless specifically asked to). DNA is nearly always referred to as just DNA, so that is acceptable in exam answers.

DNA structure and function

REVISED

DNA is made up of two long chains of alternating sugar and phosphate molecules connected by pairs of bases. This ladder-like structure is twisted to form a 'double helix' (a helix is a type of spiral). There are four bases in DNA: **adenine** (A) joins on to **thymine** (T), and **guanine** (G) joins on to **cytosine** (C). (These four names are higher tier). The order of these bases along the sugar–phosphate backbone varies in different molecules of DNA. This sequence of bases forms the instructions, in a form of code, for the manufacture of proteins. It determines which amino acids are used to make a given protein, and in what order. The structure of DNA is shown in Figure 9.1. The 'code' consists of **triplets** (groups of three) of bases along the DNA. Each triplet codes for an individual amino acid in the protein.

In the nucleus of a cell, the long DNA molecules are coiled up into structures called **chromosomes**. As we have seen, DNA is the raw material of genes – a **gene** is a short length of DNA that codes for one protein. This is summarised in Figure 9.2.

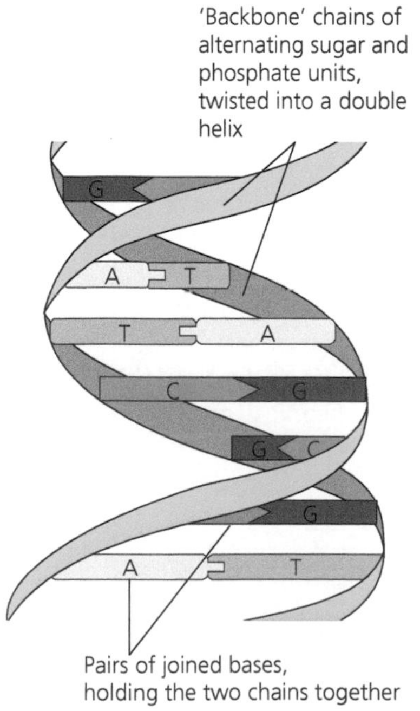

'Backbone' chains of alternating sugar and phosphate units, twisted into a double helix

Pairs of joined bases, holding the two chains together

Figure 9.1 The structure of DNA.

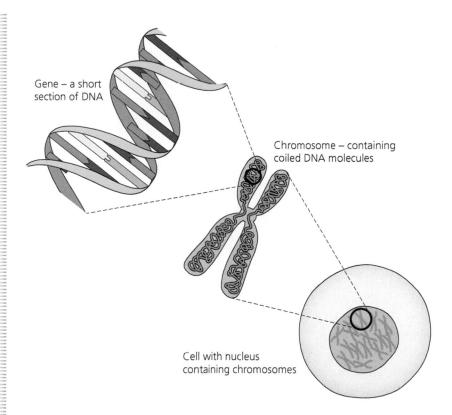

Gene – a short section of DNA

Chromosome – containing coiled DNA molecules

Cell with nucleus containing chromosomes

Figure 9.2 The structure of a gene in relation to DNA and chromosomes.

Genetic profiling and inheritance

Genetic profiling

Scientists can look at the bases in a DNA molecule and see to what extent different DNA samples are similar. The analysis of the DNA produces a **genetic profile**. The genetic profiling process can be used to show the similarity between two different samples, and some uses of this are shown below.

● Criminals can be positively identified if DNA samples from the crime scene match theirs.
● If there is a dispute as to who is the father of a child, the father's DNA will have many similarities with the child's.
● The DNA of species can be compared with each other to establish how closely related they are. The more closely related, the more similar the DNA will be.
● Genetic profiling can detect certain genes that may be associated with a particular disease. This opens up possibilities for treatment.

Genetic profiling consists of several steps:
1 A sample of cells is collected – for example, from blood, hair, semen or skin. The cells are broken up and the DNA extracted.
2 The DNA is 'cut up' by enzymes, so that it ends up in fragments of different sizes.
3 The fragments are then separated. A pattern develops, which is the 'genetic profile'.

Genetic terms

You need to know the following specialist terms used in genetics.

- **Gene** A length of DNA that codes for one protein.
- **Allele** A variety of a gene.
- **Chromosome** A length of DNA that contains many genes, found in the nucleus and visible during cell division.
- **Genotype** The genetic make-up of an individual (for example, **BB**, **Bb**, **bb**).
- **Phenotype** The description of the way the genotype 'shows itself' (for example, blue eyes, curly hair, red flowers, and so on).
- **Dominant** An allele that shows in the phenotype whenever it is present (shown by a capital letter – for example, **B**).
- **Recessive** An allele that is hidden when a dominant allele is present (shown by a lower case letter – for example, **b**).
- **F1/F2** Short for first generation (F1) and second generation (F2) in a genetic cross.
- **Homozygous/homozygote** A homozygote contains two identical alleles for the gene concerned – it is homozygous.
- **Heterozygous/heterozygote** A heterozygote contains two different alleles for the gene concerned – it is heterozygous.
- **Selfing** A technique by which pollen from a plant is used to fertilise ovules in flowers of the same plant.

> **Exam tip**
>
> It is very important to know the genetic terms on this page. Even if you are not asked directly what they mean, they will be used in questions and if you do not know the terms, you may not understand the question.

The basics of inheritance

This section links with what you learnt about cell division earlier in the book. In order to understand inheritance, you need to understand the following:

- There are two copies of each gene in the body, one coming from each parent.
- The gametes contain one copy of each gene.
- The copies are not necessarily the same, as every gene has different versions called alleles.
- Individuals may have two of the same alleles for a particular gene (homozygous) or two different alleles (heterozygous).

Monohybrid inheritance

Monohybrid inheritance is the name given to the inheritance of one gene. In this worked example, we will look at the inheritance of tall and short in peas. As you read it, check the terms given earlier to make sure you understand the points.

- Tall is dominant to short, and the tall allele is given the symbol T.
- The recessive short allele is given the symbol t.
- The dominant allele will show in the phenotype whenever it is present. So the genotypes TT (tall) and Tt will both give tall plants.
- The short phenotype will only be produced when there are two recessive short alleles. Therefore, short plants will always have the genotype tt.

In this example, we will use a cross between a homozygous tall plant (TT) and a homozygous short plan (tt) and will follow the inheritance over two generations.

1 Every gamete produced by the tall plant (TT) can only contain a T allele.

2 Every gamete produced by the short plant can only contain the t alleles.

3 Whenever pollen from one plant fertilises an egg from the other, the resulting seed will therefore contain one of each allele. It will be heterozygous (Tt). These plants are called the F1 generation.

4 When an F1 plant produces gametes, half of them will contain the T allele and half the t allele. If F1 pollen fertilises an egg from another F1 plant, there are several possible combinations. The situation described so far is summarised in Figure 9.3.

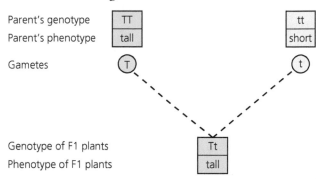

Figure 9.3 A cross between a tall and a short pea plant.

5 To work out all these possible combinations, we use something called a **Punnett square**. The gametes for each individual are put in, one set at the top and the other at the side, and then the combinations are worked out, as shown in Figure 9.4.

		Male gametes	
		T	t
Female gametes	T	TT	Tt
	t	Tt	tt

Figure 9.4 Punnett square.

You can see that in this cross, three of the possibilities produce tall plants (TT or Tt) and one produces a short plant (tt). Therefore, tall plants are three times more likely than short plants and, if a lot of offspring are produced, the ratio of tall:short plants would be approximately 3:1.

Although you can work out any given cross using a Punnet square, there are two common crosses that it is useful to remember:

● Aa × Aa gives a 3:1 ratio of dominant:recessive phenotypes.
● Aa × aa gives a 1:1 ratio of dominant:recessive phenotypes.

You study the inheritance of a single gene for simplicity, but in humans hardly any characteristic is controlled by a single gene. Our characteristics are generally controlled by the interaction of many genes.

Now test yourself

1 How could scientists tell, using genetic profiling, whether humans were more closely related to gorillas or chimpanzees?
2 Which DNA base pairs with cytosine?
3 What is a 'recessive' allele?
4 A species of plant has a gene that controls flower colour. The alleles are for red or white colour and red is dominant. Why can you not be certain of the genotype of a plant with red flowers?
5 In the species of plant in the last question a heterozygous plant (Rr) is crossed with a homozygous recessive plant (rr). Calculate the expected ratio of red:white flowers in the offspring.

Answers online

Gender and genetic modification

Gender determination

The 46 chromosomes in humans can be grouped into 23 pairs, but one of those pairs (pair 23) looks different in males and females. In females the two chromosomes look the same, but in males they are different from each other. Because of their shape, the larger chromosome is called the X chromosome and the shorter one is called the Y chromosome. Males have one X and one Y chromosome while females have two X chromosomes. This pair of chromosomes determines whether the individual is male or female.

When egg cells are formed in a female's ovary, they all contain an X chromosome (because there is no alternative), but when sperm form in a male's testis, half the sperm cells have an X chromosome and half have a Y chromosome. At fertilisation, when an egg cell fuses with a sperm cell carrying an X chromosome, a female embryo will develop, while an egg fusing with a sperm carrying a Y chromosome will develop into a male embryo. When a woman becomes pregnant, there is a 50% chance of having a boy (or a girl) as shown in Figure 9.5, because sperm and egg cells combine at random, and roughly half of the sperm cells will be carrying an X chromosome, and half a Y.

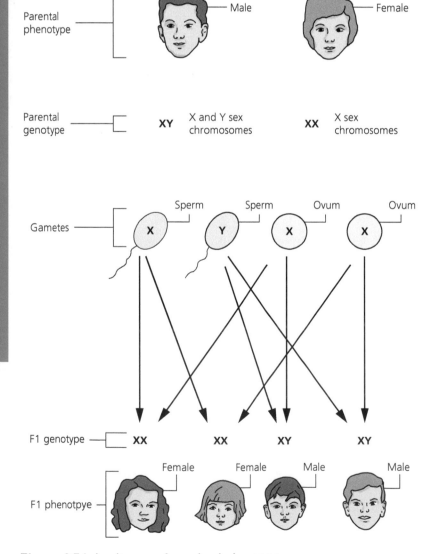

Figure 9.5 Inheritance of gender in humans.

Genetic modification

Scientists can now extract genes from one organism and put them into another, and can also 'swap' one gene for another. The introduction of genes into food plants is becoming more common and is known as **genetic modification** (sometimes shortened to GM). There are advantages, disadvantages and ethical issues involved with this technique.

Some advantages are the following:
- Crops can be 'designed' to survive in difficult farming conditions found in many of the world's poorer countries (e.g. in very hot and dry conditions).
- Crops can have a gene inserted that makes them resistant to herbicides. Using the herbicide will then only kill weeds and this could greatly increase the yield of food crops.
- Plants can be modified to produce oils or other substances that can be used as biofuels, saving non-renewable resources.

Disadvantages include:
- Sometimes the inserted gene has side effects, e.g. genetically modified soya plants had stems which split in hot conditions.
- Pollen from genetically modified crops can easily get carried out of the farm by wind or insects, and this could introduce the gene into the natural population.
- GM plants could become established outside the farm and, if they are herbicide resistant for instance, could become a pest.

Some of the ethical issues are:
- The technology is expensive and only richer countries can afford it. If they develop new 'super crops' then poorer countries' products will be unable to compete.
- The large companies which develop GM crops can patent a crop so that only they can produce it. In theory, they could then charge what they like as there will be no competition.

> **Exam tip**
>
> The specification says that you should be aware of the advantages, disadvantages and issues with GM crops, but it does not specify examples, so you just need a general awareness of these things.

Now test yourself

6 How do males and females differ in the make-up of their chromosomes?
7 In the UK the ratio of males:females 0.98:1. Explain why this is to be expected.
8 What is meant by the term genetic modification?
9 State one way in which genetic modification could benefit farmers in countries which have a challenging climate.
10 State one way in which genetic modification could harm farmers in poor countries.

Answers online

Summary

- DNA consists of two long chains of alternating sugar and phosphate molecules connected by bases; the chains are twisted to form a double helix.
- There are four types of bases – A, T, C and G – and the order of bases along the DNA molecule forms a code for making proteins; the code determines the order in which different amino acids are linked together to form different proteins.
- The full names of the bases in DNA are adenine, thymine, guanine and cytosine.
- In DNA, the base adenine (A) pairs with thymine (T), and cytosine (C) with guanine (G).
- A sequence of three bases (a triplet) determines an amino acid to be added to a protein.
- Genetic profiling involves cutting the DNA into short pieces using specific enzymes. The fragments are then separated into bands according to their size. The pattern of the bands produced can be compared to show the similarities and differences between two DNA samples – for instance, in criminal cases, paternity cases and in comparisons between species for classification purposes.
- DNA profiling can be used to identify the presence of certain genes, which may be associated with a particular disease.
- Genes are sections of DNA molecules that determine inherited characteristics.
- Genes have different forms, called alleles. In a pair of chromosomes, there may be two different alleles of a particular gene, or the two alleles may be the same.
- The following terms are used in genetics: gamete, chromosome, gene, allele, dominant, recessive, homozygous, heterozygous, genotype, phenotype, F1, F2, selfing.
- Punnett squares can be used to show the inheritance of single genes.
- The cross **Aa** × **Aa** gives a 3:1 ratio of dominant : recessive phenotypes among the offspring.
- The cross **Aa** × **aa** gives a 1:1 ratio of dominant : recessive phenotypes among the offspring.
- Most phenotypic features are the result of multiple genes rather than single gene inheritance.
- Sex determination in humans is the result of the composition of the pair of sex chromosomes; females have two X chromosomes, males have one X and one Y chromosome.
- The sex chromosomes separate in the gametes, and combine randomly at fertilisation.
- The artificial transfer of genes from one organism to another is known as genetic modification.
- Genetic modification has potential advantages and disadvantages.

Exam practice

1 The table below shows the percentages of males in the population of different countries, at different ages.

Country	% males in stated age group			
	Birth	Under 15	15–64	65+
UK	52.5	52.5	51.5	38
Sweden	53	53	51.5	39.5
Poland	53	53	49.5	31
Brazil	52.5	52	49.5	36.5
Ghana	52.5	51	50	42
New Zealand	52.5	52.5	50	42

a) Which country has the largest percentage of women at age 65+? [1]

b) Why would it be expected that the % of males at birth would be very close to 50%? [3]

c) In every country in the world the % of males born is slightly greater than 50%. Which of the following is a possible hypothesis to explain this (you may choose more than one)? [2]

i) Sperm carrying a Y chromosome swim faster than those carrying an X chromosome and so reach the egg quicker.

ii) Some egg cells have a Y chromosome.

iii) More sperm cells are produced with an X chromosome than a Y chromosome.

iv) Sperm cells carrying an X chromosome do not survive as well in the female's body as those carrying a Y chromosome.

d) What evidence does the data provide that child mortality (death of children) in the UK is not affected by gender? [1]

e) What conclusion can be drawn from the decline in the % of males in the older age groups in all countries? [1]

2 A laboratory has two strains of mice, one with black fur and the other with brown fur. They have each been bred for many generations and no other fur colours ever occur. The scientists then mated the black mice with the brown mice. All the F1 generation were black.

a) What conclusion can be drawn from this cross? [1]

b) Using the letters B or b to represent the alleles, what is the genotype of:

i) the black parent mice? [1]

ii) the brown parent mice? [1]

iii) the black F1 mice? [1]

c) Two of the black F1 mice were bred together. Over a period of time they produced 44 offspring. How many of those offspring would you expect to be brown? [2]

d) Two of the black F1 mice were interbred with two of the brown parent mice. The first cross produced a black:brown ratio of 1:1, but the second cross produced only black mice. Explain the difference. [2]

Answers and quick quiz 9 online

ONLINE

10 Variation and evolution

Variation and inherited disease

Variation

No two living things are absolutely identical. In every population of every living thing, there is variation of a huge number of features. Much of this can be accounted for by differences in the genes, but even identical twins and clones vary to some extent. Variation comes in different types, and can be grouped in two different ways.

Heritable versus environmental

Heritable variation is caused by differences in genes and so can be passed on from parent to child, e.g. hair colour, height, shape of nose etc.

Environmental variation (also known as non-heritable) is caused by the environment in which the organism has developed, e.g. scars, piercings, muscularity etc. It cannot be passed on to the next generation.

Note that some features result from a combination of heritable and environmental variation, e.g. some people are naturally tall but might not reach their potential height if they do not get a proper diet during childhood. Both genetic and environmental factors also affect weight.

Continuous versus discontinuous

In **continuous variation** there is a continuous range with no 'categories' (for example, height in humans; people can be any height within a certain range).

With **discontinuous variation** there are distinct groups (for example, fingerprints can be either an arch, a whorl or a loop – there are no 'in-between' fingerprints).

The origins of variation

Offspring are genetically different from their parents as a result of **sexual reproduction**, which involves an egg fusing with a sperm in the process of **fertilisation**. The genes from the mother in the egg are mixed with different genes from the father in the sperm. The cell formed by fertilisation (the **zygote**) has one set of genes from the father and one set from the mother. The 'set' of genes in a **gamete** represents only half of the mother's or father's total number of genes, and the combination of genes making up the 'set' varies, which is why brothers and sisters are similar but different. Organisms that reproduce by **asexual reproduction** do not mix their genes because fertilisation does not take place. One individual produces offspring that are genetically identical to each other and to the parent. These are called **clones**.

The genes of a species do not remain the same for all time. New alleles and characteristics are constantly appearing. Changes to genes are caused by **mutations**. Here are some important facts about mutations:

- A mutation is a random change in the structure of a gene.
- Mutations are very common. Ionising radiation or certain chemicals can increase the rate of mutation.
- Most mutations cause such small changes that no effect is seen.
- When a change does show, it may be harmful, or occasionally beneficial.
- Mutations in the sex cells (gametes) will be passed on to the next generation. Mutations in body cells will not.

Inherited disease

REVISED

Some mutations can result in an allele that is harmful and causes a disease. This allele, and therefore the disease, can be inherited. An example of this is **cystic fibrosis** – the lungs and digestive system of people with this disease become clogged with thick mucus, which makes breathing and digesting food difficult, and leads to reduced life expectancy.

The cystic fibrosis allele is recessive, so the disease only appears when an individual has the cystic fibrosis allele for this gene on both chromosomes. Some people have one cystic fibrosis allele and one 'normal' allele. Those people who are heterozygous for the recessive cystic fibrosis trait, will not suffer from the disease but can pass it on to any children they have. These people are called **carriers**. If two people carrying the cystic fibrosis allele have children, there is a 1 in 4 chance that any child they have will suffer from the disease. This can be shown using a Punnett square. Let us call the normal allele **C** and the cystic fibrosis allele **c**.

	mother	father
Parent genotypes:	Cc	Cc
Parent phenotypes:	normal	normal
Gametes:	C or c	C or c

		Male gametes	
		C	**c**
Female gametes	**C**	CC	Cc
	c	Cc	cc

The children with genotype CC will be clear of cystic fibrosis. Those with the genotype Cc will be carriers like their parents, and those with genotype cc will develop the disease.

Inherited diseases in family trees

REVISED

Examination of a family tree can indicate how a disease like cystic fibrosis is inherited. The cystic fibrosis allele is recessive, and with any recessive allele the only genotype that can be directly seen is the homozygous recessive. Which in this case will show cystic fibrosis? Both the homozygous dominant and the heterozygotes will be normal and so cannot be distinguished.

A worked example of an interpretation of a family tree is shown in Figure 10.1, for a family with cystic fibrosis sufferers. Interpreting the family tree, we can get some idea of the genotypes of most, but not all, of the individuals.

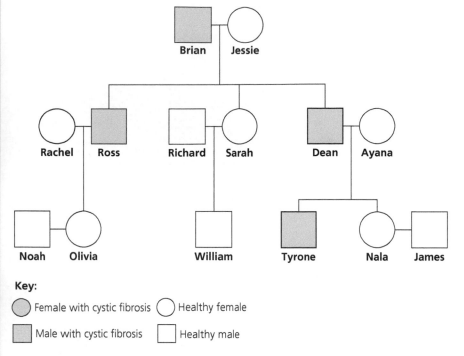

Key:

⬤ Female with cystic fibrosis ◯ Healthy female

▨ Male with cystic fibrosis ☐ Healthy male

Figure 10.1 Family tree for family including cystic fibrosis sufferers.

Let's call the normal (dominant) allele **C** and the cystic fibrosis (recessive) allele **c**.

- We can easily identify those with genotype cc, as they will have cystic fibrosis (Brian, Ross, Dean and Tyrone).
- If a child has cystic fibrosis, its parents must both be either carriers or have the disease. Jessie and Ayana must be carriers, and so have the genotype Cc.
- If a child's parent has cystic fibrosis, the child must either be a carrier or have the disease themselves. Sarah, Olivia and Nala must therefore be carriers, and so have the genotype Cc.
- We cannot identify those with the CC genotype with any certainty. Rachel, Richard and William could have this genotype, but they could also be Cc. Although Rachel and Richard have healthy children, that does not prove they are not carriers, as for a Cc × Cc cross there is only a 1 in 4 chance of having a child with cystic fibrosis, and with a Cc × cc cross the chances are 50%.
- There is no evidence at all about Noah and James, apart from the fact that they cannot be cc, as they do not have cystic fibrosis.

What we can tell about this family tree is summarised in Figure 10.2.

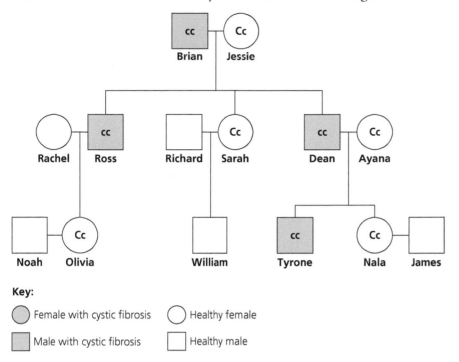

Key:

◯ (grey) Female with cystic fibrosis ◯ Healthy female

▢ (grey) Male with cystic fibrosis ▢ Healthy male

Figure 10.2 Genotypes of family members, with regard to cystic fibrosis.

Gene therapy

Gene therapy is the name given to a range of techniques that can be used to remove the effects of a harmful allele like the one causing cystic fibrosis. It can be done in two main ways:

- **Introducing a 'healthy' allele into the person's DNA** – If the harmful allele is recessive, a healthy dominant allele will counteract it. The recessive allele does not have to be removed.
- **'Switching off' the harmful allele** – This can be done in various ways, including the introduction of a completely new gene into the body.

There are ethical issues around gene therapy:

- Some religious groups believe that humans should never alter the genes of living organisms.
- The process is very expensive and could take funds from other types of healthcare which might help more people.

> **Exam tip**
>
> If you are asked about ethical objections to any process, do not say that religious groups think that people should not 'play God'. That is considered too vague – make clear exactly what is the religious objection.

Now test yourself

1 Is variation in finger length an example of continuous or discontinuous variation?
2 Is variation in finger length an example of heritable or non-heritable variation?
3 What process in cells creates variation?
4 Haemophilia is a blood disease which is inherited and caused by a recessive allele. Why is it not possible to easily distinguish carriers of the disease?
5 Tom is not a carrier of cystic fibrosis, but his partner Gwen is. What are the chances of them having a child with cystic fibrosis?

Answers online

Evolution

Evolution and variation

Evolution is the process by which living species have gradually changed and developed from earlier forms over a long period of time. Evolution results in organisms becoming better **adapted** (suited) to their environment. This can only happen because every population shows variation, which allows the more suitable variants to succeed and become the norm. How this happens is explained by the theory of natural selection, which is covered in the next section.

It is important to distinguish between the theory of evolution and the **theory of natural selection,** which are not the same.

The theory of evolution is that living things have evolved (changed) over a period of time. The theory of natural selection assumes that evolution has occurred, and is a theory of how it has happened.

Natural selection

The theory of natural selection was first proposed by Charles Darwin and Alfred Russel Wallace in 1858. It has been refined over time, but is still accepted as the mechanism of evolution by the vast majority of scientists. The theory can be outlined as follows:

- All populations vary, due to past mutations.
- Most organisms over-produce, i.e. they produce more offspring than can possibly survive. There is therefore competition for survival between the young organisms.
- If some of the young have variations that allow them to survive better in their environment, they are more likely to survive than their rivals.
- The organisms that survive are the ones who will breed, and pass on their genes to the next generation.
- The next generation will have more of the beneficial variations than the previous generation, but will still vary. Once again, the ones with the most suitable characteristics will survive, so the process repeats, generation after generation.
- The result is that the beneficial characteristics become more and more common, eventually spreading to the whole population, which will have changed, or evolved, as a result.

> **Exam tip**
>
> A common mistake in exams is to say (or imply) that mutations happen when an organism reaches a new environment or when an environment changes. The mutations that may suit an organism to a new environment will have happened long before.

Modelling natural selection

Scientists sometimes use artificial 'models' to mimic real life. One such model uses coloured objects (e.g. cocktail sticks) to model the effect of camouflage on predation, to simulate natural selection. The cocktail sticks are of two different colours and are spread on a coloured background which is the same as one of the colours of the sticks. The sticks that are the same colour as the background are camouflaged:

- The cocktail sticks are spread on the coloured background.
- Volunteers are asked to pick up a certain number of cocktail sticks in a short period of time. The volunteer acts as a model predator.
- The sticks that have not been picked up then 'reproduce', i.e. for each cocktail stick, another of the same colour is added. This is repeated many times (i.e. many generations) and the numbers of the two colours recorded each time.

> **Exam tip**
>
> You do not need to know details of the technique used here involving cocktail sticks, which is only given for clarification. You do need to be able to recognise the weaknesses in such a model of predation and natural selection, however.

Eventually, all or nearly all the remaining sticks will be of the camouflaged colour, showing that natural selection has taken place.

This model is basic and is inaccurate in many ways, for example:
- The differences in colour are usually fairly extreme, whereas in nature they would be much more similar.
- In nature, other factors will also affect survival and reproduction.
- Each surviving stick reproduces one new stick. Animals and plants usually produce much more offspring than that.
- Many organisms cannot reproduce asexually, and so the idea of each cocktail stick producing another one is not accurate for them. One organism could not reproduce on its own.
- Natural environments are not one, plain colour.

Extinction

REVISED

Millions of species that existed in the past are no longer found on Earth – they have become **extinct**. This could happen for a variety of reasons:
- The organism has failed to adapt quickly enough to its environment.
- The organism has adapted to its environment to some extent, but another similar organism has adapted better. The less successful organism cannot compete and eventually dies out.
- The organism has adapted to its environment well, but the environment suddenly changes and the organism cannot survive in the new conditions.

Evolution is a slow process and it is very difficult for populations to adapt to sudden drastic changes. Such changes rarely occur naturally, and nearly always result from human interference.

Natural selection and superbugs

REVISED

Natural selection is a constant and ongoing process. It usually takes a very long time, but in certain circumstances it can happen quite quickly. One example is the evolution of 'superbugs'. These are bacteria that have become resistant to the antibiotics normally used to treat infections. Bacteria show variation like every living organism. Most of them are susceptible to antibiotics, but when you take these medicines there will always be a few bacteria that are naturally resistant and will survive. If there are only a few of them in you, they will not cause problems, but they could still be spread to others. If lots of people use the same antibiotics, after a while the susceptible bacteria are mostly killed off, leaving the resistant ones able to multiply. Given time, the whole population of bacteria will be resistant to the antibiotic (Figure 10.3). The bacteria have not become a new species but in a fairly short period of time they have evolved to be resistant to certain antibiotics by a process of natural selection.

Natural selection is rapid in bacteria because:
- They reproduce very rapidly (about once every 20 minutes, on average), and in the course of a day they can progress through 72 generations.
- The level of exposure to antibiotics is very high because a huge number of antibiotics are used.

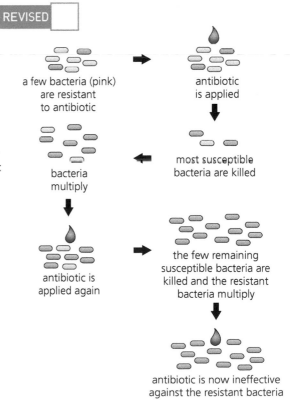

a few bacteria (pink) are resistant to antibiotic

antibiotic is applied

bacteria multiply

most susceptible bacteria are killed

antibiotic is applied again

the few remaining susceptible bacteria are killed and the resistant bacteria multiply

antibiotic is now ineffective against the resistant bacteria

Figure 10.3 Evolution of resistance to antibiotics in bacteria. Note that in reality many more generations would be required before full resistance evolved.

To avoid resistance building to new antibiotics, doctors try not to prescribe them unless it is absolutely essential, and they try to use a variety of different antibiotics.

A similar problem has been encountered with some widely used pesticides and the rat poison warfarin, which have killed off most of the pests that are susceptible to it, leaving those that are naturally resistant to it to breed and spread.

Mapping the human genome

The **genome** is the name given to all the genetic information in an organism. It includes all the genes and their sequence on all the chromosomes, and the DNA base pairs that make up those genes.

The Human Genome Project was an international scientific research project that worked out the sequence of chemical base pairs in human DNA, identified all the genes (and their variants) in humans and their location on the chromosomes. It was completed in 2003.

Mapping the human genome was an enormous task and it has huge potential importance for medicine in the future.
- Some genes are known to directly cause disease – for example, there are mutations of certain genes that can cause cancer. Others can make a person more susceptible to a disease, while others play a key role in protecting the body. Knowing about the existence of these genes and their location on the human chromosomes allows the possibility of altering them or counteracting their effects.
- It allows the possibility of creating targeted drugs or viruses that would (for example) only attack cells containing a mutated cancer-causing gene.

There appear to be about 20 500 genes in a human being (which is fewer than was previously thought), and many of them interact in complex ways. Mapping the genome is just the start and work is ongoing to discover exactly how these genes work, which could lead to further developments and applications.

Now test yourself

TESTED

6 Which two scientists were responsible for developing the theory of natural selection?
7 Why are bacteria like 'superbugs' able to evolve quicker than other organisms?
8 State one way in which doctors are trying to slow down the evolution of superbugs.
9 What is a genome?

Answers online

Summary

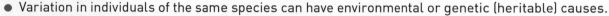

- Variation in individuals of the same species can have environmental or genetic (heritable) causes.
H - Variation can be classified as continuous or discontinuous.
- Sexual reproduction leads to offspring being genetically different from the parents, but in asexual reproduction the offspring are genetically identical clones.
- New alleles result from mutations, which are changes in existing genes; mutations occur at random and often have no effect, but some can be beneficial or harmful.
- Mutation rates can be increased by environmental factors such as ionising radiation.
- Some mutations cause inherited diseases such as cystic fibrosis.
- Gene therapy offers hope to cystic fibrosis sufferers and those with other heritable diseases.
- Heritable variation is the basis of evolution.
- Individuals with characteristics adapted to their environment are more likely to survive and breed successfully. Genes that have enabled these better adapted individuals to survive are then passed on to the next generation.
- The theory of natural selection was proposed by Alfred Russel Wallace and Charles Darwin.
- The process of natural selection is sometimes too slow for organisms to adapt to rapidly changing environmental conditions and so species may become extinct.
- Antibiotic resistance in bacteria, pesticide resistance and warfarin resistance in rats are examples of natural selection happening over a shorter than usual period of time.
- Understanding the human genome opens up many potential benefits in medicine, allowing new and targeted forms of treatment.

Exam practice

1 Malaria is a disease carried by mosquitos. In the middle of the 20th century, the insecticide DDT was used to try to kill mosquitos in eastern Africa. A very large amount of DDT was used across a large area, and as a result, malaria was almost eradicated. However, by the late 20th century malaria was once again a problem in the area. Explain how natural selection can explain this sequence of events [6]

2 The family tree below shows the incidence of cystic fibrosis in one family.

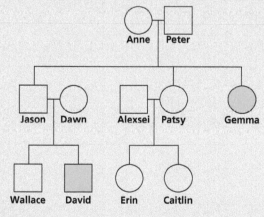

Key:
⬤ Female with cystic fibrosis ◯ Healthy female
⬛ Male with cystic fibrosis ☐ Healthy male

a) What are the genotypes of the following family members? [3]
    i) Gemma
    ii) Peter
    iii) Dawn
b) Can Alexsei and Patsy be certain that if they have more children, none will be born with cystic fibrosis? Explain your answer. [5]
c) Dawn is pregnant. What are the chances that her third child will have cystic fibrosis? [1]

Answers and quick quiz 10 online

ONLINE ☐

11 Response and regulation

The nervous system and coordination

The nervous system

REVISED

The nervous system controls the activity of the body in the following ways:
- It detects changes in the environment.
- It controls the actions of the body.
- It coordinates the responses of the body by linking various actions in specific ways, and making decisions about whether any action is needed, and what sort of action it should be.

Sense organs

REVISED

Information is fed to the brain constantly by a system of **sense organs** scattered around the body (Figure 11.1).Sense organs are groups of special cells called **receptor cells**, which can detect changes around them, either internally or in the external environment. These changes are called **stimuli** (singular: stimulus) and include light, sound, chemicals, touch and temperature. Each group of receptor cells responds to a specific stimulus. The information from the sense organs travels to the brain and spinal cord (the **central nervous system**) along nerve cells (also called **neurones**) which are grouped together to form **nerves**. The brain coordinates the information from the sense organs and takes appropriate action, if necessary.

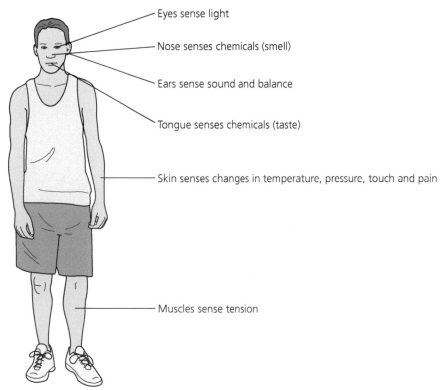

Eyes sense light

Nose senses chemicals (smell)

Ears sense sound and balance

Tongue senses chemicals (taste)

Skin senses changes in temperature, pressure, touch and pain

Muscles sense tension

Figure 11.1 Some of the body's sense organs.

Structure of the nervous system

The parts of the nervous system are shown in Figure 11.2.

Your brain and spinal cord make up the **central nervous system (CNS)**. Together they coordinate and control your body – to do this, they need to receive information from the sense organs, and send out information to muscles to make a **response** happen. Together, the central nervous system and the nerves form the **nervous system**. Some other important facts about the nervous system are:

- The information travels along the neurones as an electrical current, called an **impulse**.
- Some neurones carry impulses from the brain to muscles and glands and make something happen. These are called **motor neurones**.
- Other neurones send information from the sense organs to the brain and spinal cord. These are called **sensory neurones**.
- When a response happens, the time taken between the stimulus and the response is known as the **reaction time**.

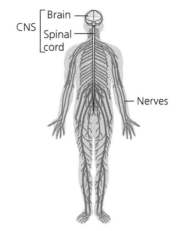

Figure 11.2 The human nervous system.

> **Exam tip**
>
> When talking about information travelling along nerves, always refer to it as an impulse, not a 'signal', which is considered to be a bit vague.

Reflexes

A reflex is a particular type of response to a stimulus. The characteristics of a reflex are:

- It is a very rapid response.
- It is automatic, i.e. involuntary.
- It is protective in some way.

Examples of reflexes include:

- The 'knee jerk' reflex
- Breathing
- Blinking
- Swallowing
- Sneezing
- Coughing
- Pupil reflex (your pupil getting bigger in the light and smaller in the dark)
- The 'withdrawal reflex' (pulling away from a painful stimulus).

Some of these things can be done voluntarily (e.g. blinking, coughing) but in that case they would not be classed as a reflex.

A reflex involves a stimulus, a receptor, a coordinator, an effector and a response:

- A **stimulus** is a change in the environment that can be detected.
- A **receptor** is an organ that detects the stimulus.
- A **coordinator** detects the signal from a receptor and sends an impulse to the effector.
- The **effector** is the part of the body (usually a muscle) that produces the response.
- The **response** is the action carried out.

The way the withdrawal reflex works is shown in Figure 11.2.

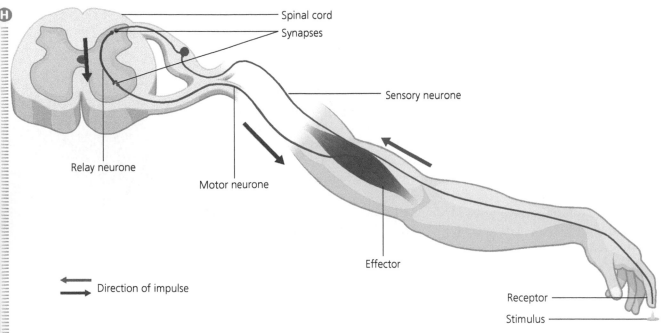

Spinal cord
Synapses
Sensory neurone
Relay neurone
Motor neurone
Effector
Receptor
Stimulus

Direction of impulse

Figure 11.3 A reflex arc for the withdrawal reflex.

The stages in a reflex are as follows:
1 The stimulus is received by the **receptor**.
2 An impulse is sent along the **sensory neurone** to the spinal cord.
3 The impulse moves across a tiny gap (**synapse**) to the **relay neurone**.
4 The relay neurone transmits the signal (via another synapse) to the **motor neurone**.
5 The motor neurone stimulates the **effector** (the muscle) to respond.

The impulse is automatic because it does not go through the brain, which is the part of the nervous system that 'makes decisions'. The impulse is quick because it is the synapses that slow down an impulse, and in a reflex the impulse only goes through two. If it went through the brain, it would go through thousands of synapses.

Now test yourself

TESTED

1 What name is given to nerve cells?
2 What is reaction time?
3 Why is a reflex involuntary?
4 In a reflex, what name is given to the part of the body that carries out the response?
5 What name is given to the small gap between neurones?

Answers online

Homeostasis

Homeostasis and control

REVISED

Homeostasis is the name given to the collection of processes which keep certain conditions in the body at a constant level. This maintains optimum conditions for the chemical reactions going on in cells.

Hormones play a vital role in homeostasis. They are chemical messengers that are made in certain organs and travel around in the bloodstream, affecting various specific parts of the body. They are mainly used for medium-term and long-term regulation, whereas nerves generally control quicker responses.

The main conditions that are controlled by hormones in the body are described below:

- **Temperature.** This must be kept stable at the optimum level for enzymes to work.
- **Water content.** Too little water (dehydration) may make the body fluids too concentrated and damage the body, but too much water can also be dangerous, as it dilutes the body fluids. The concentration of our bodily fluids is maintained within safe limits by hormones.
- **Glucose concentration.** Glucose is essential for energy but if the level gets too high it can damage the body. For this reason it is essential that glucose levels are kept within a safe range, neither too low nor too high.

Control of blood glucose

REVISED

When blood glucose levels are raised following a meal, they can be reduced by the hormone **insulin** (a protein), which is released by the pancreas. Insulin is released into the bloodstream, where it converts soluble glucose to an insoluble carbohydrate called **glycogen**, which is stored in the liver.

Some people have a condition that means their body produces little or no insulin. If untreated, their blood glucose levels become dangerously high. This condition is called **diabetes**. Activities also influence blood glucose levels. Eating carbohydrates raises blood glucose, while exercise lowers it.

Diabetes

REVISED

There are two types of diabetes, called type 1 and type 2.

In **type 1 diabetes** (the most common type in young people), the body stops producing insulin. It is thought that type 1 diabetes may be brought on by the body over-reacting to a certain type of virus, with the result that the immune system destroys its own insulin-producing cells in the pancreas. As a result, blood glucose levels go up and up and the body tries to get rid of the excess glucose in the urine. If diabetes is not treated, the blood sugar level will become so high that the person will die. It cannot be cured but it can be managed so that the sufferer remains otherwise healthy. The treatment of type 1 diabetes consists of three things:

- The person has to inject themselves with insulin (usually before every meal) to replace the natural insulin that is no longer being produced.
- The diet has to be carefully managed. The patient has to eat the right amount of carbohydrate (which is the source of glucose) to match the amount of insulin injected.
- The patient usually tests his or her blood glucose levels several times a day, to make sure the level has not gone too high or too low.

Type 2 diabetes is more common in older people. It is not caused by a lack of insulin, but the body no longer responds properly to the insulin produced. It is milder and can usually be controlled by drugs, such as metformin tablets, or even by just being careful with the diet. Type 2 diabetes tends to be associated with being overweight or obese.

⊕ Negative feedback

REVISED

Negative feedback is a control mechanism often used in living organisms to keep a particular factor within an acceptable range. It is a mechanism whereby a change in a factor sets off a series of events that lead to that factor being brought back to the normal level.

H An example of negative feedback is the control of blood glucose levels. A rise in blood glucose sets in motion a series of events that results in the level being lowered again. In the same way, low blood glucose causes a process that raises the level. This mechanism is an example of **negative feedback**, and is summarised in Figure 11.4. It involves insulin and another hormone from the pancreas, **glucagon**, which raises blood glucose.

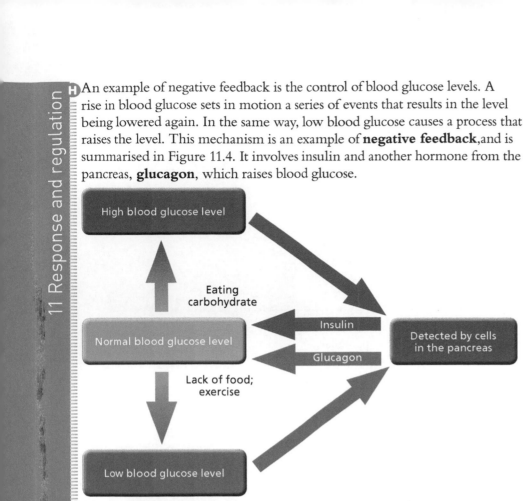

Figure 11.4 Negative feedback control of glucose levels.

Temperature control

REVISED

Animals are kept alive by a series of chemical reactions that take place in their cells. These reactions are controlled by enzymes, which are affected by temperature. If the body temperature is not kept constant, essential reactions could stop. Mammals and birds control their body temperature precisely using various mechanisms. In humans, the body temperature is maintained at approximately 37 °C.

The skin is the main organ that controls temperature, and its structure is shown in Figure 11.5.

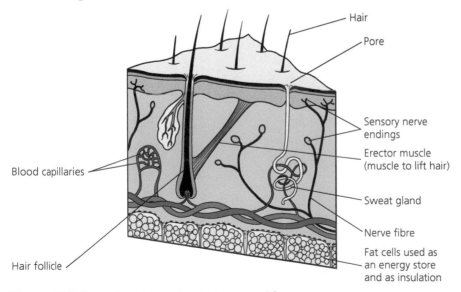

Figure 11.5 A section through the human skin.

The skin controls the body temperature in the following ways:

When it is hot:

- The **sweat glands** produce a lot of sweat. The evaporation of the sweat draws heat out of the skin to change the liquid water into vapour.
- The blood vessels near the surface of the skin **dilate** (get wider) and more blood flows close to the atmosphere. This allows heat to pass out of the blood into the cooler air.
- The muscles attached to the hair relax to flatten the hair, reducing the insulating effect of the layer of air trapped between the hair.

When it is cold:

- The sweat glands produce less sweat.
- The blood vessels near the surface **constrict** (get narrower). Less blood flows through them so less heat is lost to the air.
- The muscles attached to the hair contract so that the hairs stand on end. The layer of air trapped between the hair acts as insulation and reduces heat loss.

In cold conditions, in addition to the actions of the skin, the body shivers. This contraction of muscles generates heat.

> **Exam tip**
>
> In questions on temperature control, a common mistake is to say that blood vessels move closer to the surface or further away. The vessels cannot move – they simply channel more or less blood near to the surface by dilating or constricting.

Homeostasis and lifestyle

REVISED

We have already seen that lifestyle choices can affect the risk of developing type 2 diabetes. The human body is incredibly complex and anything that disrupts any of the chemical reactions that makeup life can have health effects. In type 2 diabetes the mechanism is not yet clear, but it seems that being overweight somehow alters the body chemistry and results in cells being less able to respond to insulin. Drugs have multiple effects on the chemical reactions in the body. A drug is defined as a substance that alters the way the body works. While some of these changes can be beneficial over the short term, long-term use or abuse can create damaging side effects. Alcohol and the nicotine in tobacco are drugs, and their effects on body chemistry are what make them harmful to health, too.

> **Now test yourself**
>
> TESTED
>
> 6 Which chemical forms a store of glucose in the liver?
> 7 What name is given to the hormone that has the opposite action to insulin?
> 8 Which type of diabetes is related to lifestyle?
> 9 How does sweating help to lower the body temperature?
> 10 Why does the skin of people appear red when they are hot?
>
> Answers online

Summary

- Sense organs are groups of receptor cells that respond to specific stimuli.
- The sense organs relay information as electrical impulses along neurones to the central nervous system.
- The brain, spinal cord and nerves form the nervous system; the central nervous system consists of the brain and spinal cord.
- Reflex actions are fast and automatic and some are protective.
- A reflex always has the following components: stimulus, receptor, coordinator and effector.
- A reflex arc consists of: a receptor, a sensory neurone, a relay neurone in the spinal cord, a motor neurone, an effector and synapses.
- Plant shoots respond to light (phototropism) and plant roots to gravity (gravitropism).
- Phototropism is due to a plant hormone, called auxin.
- The eye contains the following structures: sclera, cornea, pupil, iris, lens, choroid, retina, blind spot and optic nerve. Each of these plays a specific part in vision.
- Animals need to regulate the conditions inside their bodies to keep them relatively constant.
- Hormones are chemical messengers, carried by the blood, which control many body functions

- In the human body, glucose levels need to be kept within a constant range.
- When blood glucose levels rise, the pancreas releases the hormone insulin (a protein) into the blood, which causes the liver to reduce the glucose level by converting glucose to insoluble glycogen and then storing it.
- Diabetes is a common disease in which a person is not able to adequately control blood glucose level.
- Type 1 diabetes is caused by the body not producing insulin. It is treated by injections of insulin.
- Type 2 diabetes is caused by the body cells not properly responding to the insulin that is produced. It is treated by controlling the diet and by tablets.
- The following processes contribute to the control of body temperature: change in diameter of blood vessels near the skin, sweating, erection of hairs, shivering.
- Negative feedback mechanisms maintain optimum conditions inside the body.
- Some conditions are affected by lifestyle choices, such as poor diet, over-consumption of alcohol and drug abuse. These affect the chemical processes in people's bodies.

Exam practice

1 The graph shows the body temperatures of two athletes running for a period of two hours, one taking regular fluids and the other taking none. A body temperature of over 40 °C is considered dangerous.

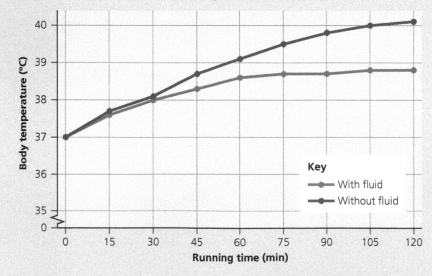

a) What is the latest time that the athlete without fluids should have been stopped? [1]

b) What is the difference in body temperatures of the two athletes after 1 hour? [1]

→

c) Explain why the athlete who took no fluids was unable to control his body temperature properly, but the one taking fluids was. [2]

d) Suggest a reason why there was very little difference in the body temperatures of the two athletes during the first 15 minutes. [1]

e) The skin temperature is normally around 33 °C. Why is it more difficult to control body temperature when the environmental temperature is 34 °C or above. [2]

f) In hot temperatures, people feel much warmer and more uncomfortable in areas where the air is humid compared to where it is drier. Suggest the reason for this. [2]

2 James did a reaction time test which tested his reaction time to a shape appearing on a screen. The test was repeated 10 times. The first time he did the test using his right hand, and on the second occasion he used his left. James is right handed. The results are shown below.

Test no.	Reaction time (ms)	
	Right hand	Left hand
1	328	340
2	309	336
3	310	320
4	287	301
5	289	298
6	268	282
7	277	260
8	274	279
9	281	282
10	270	272
Mean	289.3	

a) Define the term 'reaction time'. [1]
b) Calculate the mean reaction time for the left hand. [1]
c) Describe the trend seen in these results and suggest a reason for that trend. [2]
d) It would be incorrect to conclude that James reacted faster with his right hand. Suggest one weakness in the evidence for a difference between the right and left hand results. [1]
e) Why would it be incorrect to describe James' response to the shape as a 'reflex'? [1]
f) What type of nerve cell would carry the impulse to James' muscles when he reacted to the shape? [1]

Answers and quick quiz 11 online

ONLINE

12 Microorganisms and disease

Basics of microbiology

Types of microorganism

A microorganism is a living organism which is so small that it can only be seen with a microscope. There are four types:
- Viruses
- Bacteria
- Fungi
- Protists.

Fungi are the odd one out here, as not all fungi are microorganisms.

Most microorganisms are harmless and some are beneficial (e.g. gut bacteria help with digestion; soil bacteria break down the dead bodies of plants and animals). Some microorganisms cause disease and those are referred to as pathogens.

> **Protist:** An organism belonging to the Kingdom Protista. Many protists consist of just one cell, which is eukaryotic (that is, contains a nucleus).

Structure of bacteria

Bacteria are one celled organisms, but their cell structure is different to animal and plant cells. Figure 12.1 shows the features which can be found in bacterial cells, and highlights the differences between them and other cells.

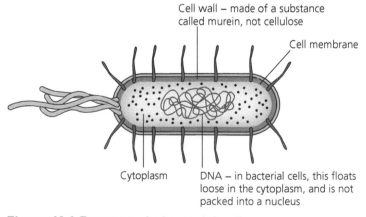

Cell wall – made of a substance called murein, not cellulose

Cell membrane

Cytoplasm

DNA – in bacterial cells, this floats loose in the cytoplasm, and is not packed into a nucleus

Figure 12.1 Features of a bacterial cell.

Structure of viruses

The structure of a virus is so different from the cells we have seen so far that it cannot really be called a cell at all. The structure of the influenza (flu) virus is shown in Figure 12.2.

A virus is really just some genes in a protein coat. There is no cytoplasm and no cell membrane.

Viruses are even smaller than bacteria.

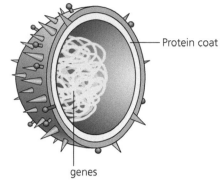

Figure 12.2 Structure of the influenza virus, shown in cross-section.

Growing microorganisms

Microorganisms are very small and difficult to see. To study them, scientists grow them into large numbers, which they can work with. This is often done on a special type of jelly called **agar**, which has nutrients added to feed the microorganisms, in a special plate called a **Petri dish**. Bacteria grow very quickly indeed, and each bacterium that lands on the agar will soon grow into a circular patch called a **colony**, which can be seen with the naked eye.

Because each bacterium grows into a colony, by counting the colonies we know how many bacteria were originally put onto the plate in the original sample.

Bacterial populations grow incredibly quickly, but the growth rate is dependent on temperature. Warm temperatures are optimum for growth. At low temperatures, the rate of growth is slower and at very high temperatures, the bacteria are killed. We put fresh food in refrigerators because the cool temperature means that the bacteria will grow slowly, and so the food will last longer before going off. To keep it even longer, we put it in a freezer, where the temperature is so cold it virtually stops bacterial growth altogether. Freezing food does not kill the bacteria, however – it just stops them growing.

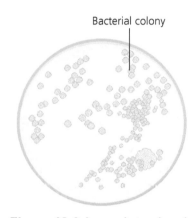

Figure 12.3 Agar plate showing bacterial colonies.

Now test yourself

1 What is the difference between a plant cell wall and a bacterial cell wall?
2 What is the cell coat of a virus made of?
3 5 cm³ of bacterial culture was spread on an agar plate and incubated. After 24 hours, 295 colonies were seen. How many bacteria were there in 1 cm³ of the culture?
4 Why is it dangerous to defrost food from a freezer, and then put the unused food back into the freezer?
5 Why does food keep longer in a freezer than in a refrigerator?

Answers online

The immune system

The immune system provides two lines of defence, first preventing the entry of pathogens, then, if that fails, killing them inside the body. Preventing entry consists of two features.

- The skin is an impenetrable barrier against organisms. Most of the body is covered in skin, apart from body openings.
- If the skin is broken then the blood at the site of the wound clots, sealing the gap.

If they get inside the body, pathogens are killed by white blood cells. There are two types:

- **Phagocytes**, which ingest ('eat') microorganisms and digest them.
- **Lymphocytes**, which produce chemicals called **antibodies** which destroy microorganisms and others called **antitoxins** which neutralise any poisons produced by the pathogen.

Now test yourself

TESTED

6 Explain how clotting prevents infection.

Answers online

Antigens, antibodies and vaccination

Antigens and antibodies

REVISED

The white blood cells can detect that cells are 'foreign' and do not belong to the body by recognising certain molecules on their cell surface. These molecules are called **antigens**, and all the cells of your body have identical antigens. If the white blood cells come across any cell that does not have the 'correct' pattern of antigens, they know it is an invader, and attack it.

The lymphocytes respond to foreign antigens by producing chemicals called **antibodies**. Phagocytes will attack any foreign cell, but the lymphocytes' response is specific – the antibodies they produce will depend on the antigens detected. The antibodies can either destroy the microorganisms, or may stick them together so that the phagocytes can ingest a lot at once.

The body's defence system works like this:
1 The antigens are detected by the white cells.
2 The phagocytes attack, and the lymphocytes start to develop an antibody against the antigens. The development of these antibodies takes some time, and during this time the numbers of pathogens will grow, possibly causing disease symptoms.
3 When the lymphocytes start to produce the right antibody, they also form **memory cells**, which will be able to produce the correct antibody instantly if the pathogen gets into the body again.
4 The antibodies will start to work against the pathogen and in most cases will wipe it out.
H 5 The formation of memory cells means that if the same pathogen gets into the body again, the response can be more rapid and the pathogen will be eliminated before symptoms occur. The person will have become **immune** to that specific disease.

Vaccination

Many diseases can be prevented by being **vaccinated** before you encounter the pathogen. Vaccines can protect against both bacterial and viral diseases.

When you are vaccinated against a disease, you are injected with the microorganisms that cause the disease. However, the pathogen has either been killed or (less often) weakened so that it is incapable of causing symptoms. It still has its antigens, so the lymphocytes can react to it and build up memory cells and therefore the body becomes immune to that disease. The microorganisms do not reproduce inside the body, and so with a vaccine the immune response is not as great as it is when you get the disease. To build up enough memory cells to give full immunity, one or more 'booster' injections are sometimes needed after the first vaccination.

Vaccines have to be given before you encounter the pathogen, so they tend to be given to children. The decision to vaccinate children is made by parents. This is not always an easy decision, even though it will give the child immunity, for several reasons:
- The vaccination often involves an injection, which can hurt and can scare young children.
- Vaccinations usually have some side effects. These are minor (for example, itching and inflammation of the injection site or feeling unwell for a day or so) but in the past some scare stories have put parents off the idea of vaccinations.

Superbugs

The immune system can sometimes be helped out by the use of **antibiotics**, which are chemicals which kill bacteria (antibiotics are useless against viruses). Recently, certain species of bacteria have evolved resistance to antibiotics. Doctors now use a large variety of antibiotics and resistance to one or a few of them would pose no major problem, but some bacteria have evolved resistance to most of the antibiotics used, to become so-called 'superbugs'. One example is MRSA (methicillin-resistant *Staphylococcus aureus*). This can cause life-threatening conditions such as blood poisoning.

Natural selection in the bacteria has caused the evolution of antibiotic resistance, as described earlier in this book. The rate of natural selection has been increased by the extensive use of antibiotics and, although there are still a few antibiotics that can treat MRSA infections, it seems that the bacterium is evolving resistance faster than we can produce new antibiotics.

The control measures for MRSA fall into two categories – the prevention of infection and combating the evolution of resistance.

To prevent infection, the following measures are adopted:
- Patients entering hospital are screened for MRSA.
- Hospital staff are rigorous with personal hygiene, washing their hands before carrying out procedures.
- Visitors are recommended to wash their hands or use hand sanitiser gels when entering wards.
- Stringent hygiene measures are taken with any procedure involving body openings or wounds.

To slow natural selection, the following measures are adopted:
- Doctors avoid prescribing antibiotics wherever possible – for example, if an infection is mild and the body can overcome it without antibiotics.

● Where antibiotics are prescribed, doctors vary the type as much as possible. Extensive use of any single antibiotic increases the risk that it will become ineffective.

Summary

● Most microorganisms are harmless and many are beneficial; some microorganisms, called pathogens, cause disease.
● Pathogens include bacteria, viruses, protists and fungi.
● The basic structures of a bacterial cell and a virus are described.
● The body defends itself from disease by having an intact skin that forms a barrier against microorganisms, blood clots to seal wounds, phagocytes in the blood to ingest microorganisms and lymphocytes to produce antibodies and antitoxins.
● An antigen is a molecule that is recognised by the immune system – foreign antigens trigger a response by lymphocytes, which secrete antibodies specific to the antigens.
● Antibodies can kill the microorganism concerned or assist the phagocytes' ability to engulf them.
● Vaccination can be used to protect humans from infectious disease.
● Certain factors have influenced, and continue to influence, parents' decisions about whether to have children vaccinated or not.

(H)
● A vaccine contains antigens derived from a disease-causing organism.
● A vaccine will protect against infection by that organism by stimulating the lymphocytes to produce antibodies to that antigen.
● Vaccines can protect against diseases caused both by bacteria and by viruses.
● Once an antigen has been encountered, memory cells remain in the body and antibodies are produced very quickly if the same antigen is encountered a second time – this memory provides immunity following a natural infection and after vaccination but is specific to one microorganism.
● Antibiotics, including penicillin, were originally medicines produced by living organisms, such as fungi.
● Antibiotics help to cure bacterial disease by killing the infecting bacteria or preventing their growth, but do not kill viruses.
● Some resistant bacteria, such as MRSA, can result from the overuse of antibiotics – methods have been introduced to try to control MRSA.
● Some conditions can be prevented by treatment with drugs or by other therapies.

Exam practice

1 Cattle were given a vaccine against ringworm, a fungal infection. The levels of antibodies in their blood were monitored over a period which included a booster vaccine. The results are shown below.

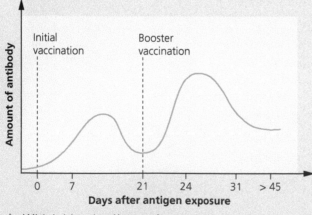

a) Which blood cells produce antibodies? [1]
b) What term would be used to describe ringworm, as an organism which can cause disease? [1]
c) What evidence is there that the cows will need a further booster vaccine? [1]
d) The level of antibodies is greater after the booster than it was after the first vaccination. Explain why this is. [3]
e) Cattle are sometimes given antibiotics as a precaution, even though they are healthy. Explain why this practice is being discouraged in an effort to slow down the evolution of 'superbugs'. [3]

Answers and quick quiz 13 online

ONLINE

13 The nature of substances and chemical reactions

Elements and compounds

Elements

REVISED

Elements are the basic building blocks of matter. They cannot be broken into anything simpler by chemical means. The important facts about elements are as follows.

- Each element has its own symbol.
- Elements are made up of **atoms**, and all the atoms of an element are of the same type.
- Each atom contains a small positively charged central region called the **nucleus**.
- The nucleus is made up of two types of particle – **protons** (which have a positive charge) and **neutrons** (which have no charge). Light, negatively charged **electrons** surround the nucleus and are attracted to it. (Positive attracts negative.)
- Every atom of a particular element has the same number of protons, known as the **atomic number**. Each element has its own atomic number. For example, hydrogen has the atomic number 1; lithium has the atomic number 3; chlorine has the atomic number 17.
- The mass of an atom of an element is called its **relative atomic mass**. Protons and neutrons are given a mass of 1. Electrons are so small that they have a negligible mass. The nucleus contains nearly all the mass of the atom.

Chemists have arranged all the elements into a table known as the Periodic Table, which organises them into groups which allow us to understand and predict the properties of the different elements.

The Periodic Table is an essential tool for chemists, and is covered later in this book.

> **Exam tip**
>
> Knowledge and understanding of the facts about elements listed here are important in most of the topics. It is essential that you are clear on all these points and can recall them.

Chemical formulae

REVISED

All chemicals have names, but are also given chemical symbols (in the case of elements) and formulae (for molecules) as a sort of 'shorthand' way of representing them. Chemical symbols have one or two letters. Many elements have symbols that obviously relate to their name (for example, the symbol for calcium is Ca), but many common ones do not (e.g. lead – Pb; mercury – Hg; tin – Sn; potassium – K). The chemical formula for a molecule indicates the elements (or sometimes a single element) that make it up, and has a number (as a subscript) to indicate how many atoms of the element are present. If there is no number, this indicates that there is one atom of the element in the molecule. Some examples are shown in Table 13.1.

Table 13.1 Formulae of some molecules.

Name of molecules	Formula	Elements present	Number of atoms of the element
Oxygen	O$_2$	Oxygen	2
Nitric acid	HNO$_3$	Hydrogen	1
		Nitrogen	1
		Oxygen	3
Zinc sulfate	ZnSO$_4$	Zinc	1
		Sulfur	1
		Oxygen	4
Magnesium chloride	MgCl$_2$	Magnesium	1
		Chlorine	2

Compounds

A compound is a substance made of two or more elements chemically combined. The compound has completely different properties to the elements it contains. Although all compounds have names, the true nature of a compound is revealed by its formula. The chemical formula not only tells you what elements make up the compound, but also the ratio of the different atoms it contains.

There are two types of compound: **ionic** compounds (which are made up of electrically charged particles or **ions**, like the sodium and chloride ions in sodium chloride or common salt) and **covalent** compounds, which are electrically neutral.

Compounds can be represented by diagrams, showing how the atoms are joined together. An example (for water) is given in Figure 13.1. With these diagrams, it is important to either label the atoms or provide a key.

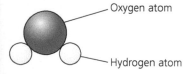
Oxygen atom

Hydrogen atom

Figure 13.1 Diagram of a water molecule.

Ionic compounds and their formulae

When metals react with non-metals ionic compounds are formed.
- **Ions** are formed from atoms when they transfer electrons.
- **Metal** atoms prefer to *lose* electrons forming *positive* ions.
- **Non-metals** prefer to *gain* electrons forming *negative* ions.
- The salts that are formed when metals react with non-metals have names that reflect the metal ion and the non-metal ion. The metal ions have the same name as their atom, so sodium atoms become sodium ions. Non-metal ions have slightly different names. Oxygen forms oxides, fluorine forms fluorides, chlorine forms chlorides, bromine forms bromides and iodine forms iodides.

The formulae of simple ionic compounds can be written using the formulae of the ions given in Tables 13.2 and 13.3. Whenever ionic compounds are formed from metals and non-metals, the resulting compounds are electrically neutral – the number of positive charges balances the number of negative charges. Sodium chloride is easy. The sodium ion is Na^+ and the chloride ion is Cl^-. The formula is $(Na^+)(Cl^-)$, which we write as NaCl. The calcium ion is Ca^{2+}. This means that two chloride ions are needed to balance the charge on one calcium ion. The formula for calcium chloride is written as $CaCl_2$.

Table 13.2 Formulae of some positive ions.

Ion	Formula
Periodic Table Group 1	
Lithium	Li^+
Sodium	Na^+
Potassium	K^+
Periodic Table Group 2	
Magnesium	Mg^{2+}
Calcium	Ca^{2+}
Strontium	Sr^{2+}

Table 13.3 Formulae of some negative ions.

Ion	Formula
Periodic Table Group 6	
Oxide	O^{2-}
Periodic Table Group 7	
Fluoride	F^-
Chloride	Cl^-
Bromide	Br^-
Iodide	I^-

The formulae of other compounds can also be written from the formulae of their ions. The formulae of some common ions will be given in the examinations (at the back of the exam paper) as shown in Tables 13.4 and 13.5.

Table 13.4 Formulae for positive ions.

Charge: +1		Charge: +2		Charge: +3	
Sodium	Na^+	Magnesium	Mg^{2+}	Aluminium	Al^{3+}
Potassium	K^+	Calcium	Ca^{2+}	Iron(III)	Fe^{3+}
Lithium	Li^+	Barium	Ba^{2+}	Chromium(III)	Cr^{3+}
Ammonium	NH_4^+	Copper(II)	Cu^{2+}		
Silver	Ag^+	Lead(II)	Pb^{2+}		
		Iron(II)	Fe^{2+}		

Table 13.5 Formulae for negative ions.

Charge: –1		Charge: –2		Charge: –3	
Chloride	Cl^-	Oxide	O^{2-}	Phosphate	PO_4^{3-}
Bromide	Br^-	Sulfate	SO_4^{2-}		
Iodide	I^-	Carbonate	CO_3^{2-}		
Hydroxide	OH^-				
Nitrate	NO_3^-				

Now test yourself

TESTED

1 What are the two types of particle found in an atomic nucleus?
2 The element cobalt has 27 protons and 32 neutrons. What is its atomic number?
3 The formula for sulfuric acid is H_2SO_4. How many atoms of hydrogen, sulfur and oxygen are in a molecule of sulfuric acid?
4 What type of substance forms positive ions?
5 Look at Tables 13.4 and 13.5. What would be the formula for: a) Potassium hydroxide? b) Sodium phosphate? c) Magnesium oxide? d) Calcium chloride?

Answers online

Chemical data and percentage composition of compounds

Relative atomic mass and relative formula mass

REVISED

If you try to express the weight of an atom in grams, the number would be incredibly small. To have figures that are easier to use, scientists express the mass of an atom relative to the mass of the carbon atom, ^{12}C. The mass of ^{12}C is given as 12 and the masses of other atoms are then calculated from that. These figures are called **relative atomic masses**, and given the symbol A_r. The relative atomic masses of some common elements are given in Table 13.6. The relative atomic mass is often given, along with the atomic number, in a Periodic Table (see Figure 13.2). A_r is a comparison, not an actual measurement, so it has no units.

Table 13.6 A_r values of some common elements.

Element	A_r	Element	A_r	Element	A_r
H	1.0	Na	23.0	K	39.0
He	4.0	Mg	24.0	Ca	40.0
C	12.0	Al	27.0	Fe	56.0
N	14.0	P	31.0	Cu	63.5
O	16.0	S	32.0	Ag	108.0
F	19.0	Cl	35.5	Pb	207.0

Group 1	Group 2											Group 3	Group 4	Group 5	Group 6	Group 7	Group 0
						1 H 1											4 He 2
7 Li 3	9 Be 4											11 B 5	12 C 6	14 N 7	16 O 8	19 F 9	20 Ne 10
23 Na 11	24 Mg 12											27 Al 13	28 Si 14	31 P 15	32 S 16	35.5 Cl 17	40 Ar 18
39 K 19	40 Ca 20	45 Sc 21	48 Ti 22	51 V 23	52 Cr 24	55 Mn 25	56 Fe 26	59 Co 27	59 Ni 28	63.5 Cu 29	65 Zn 30	70 Ga 31	73 Ge 32	75 As 33	79 Se 34	80 Br 35	84 Kr 36
85 Rb 37	88 Sr 38	89 Y 39	91 Zr 40	93 Nb 41	96 Mo 42	98 Tc 43	101 Ru 44	103 Rh 45	106 Pd 46	108 Ag 47	112 Cd 48	115 In 49	119 Sn 50	122 Sb 51	128 Te 52	127 I 53	131 Xe 54
133 Cs 55	137 Ba 56	139 La 57	178 Hf 72	181 Ta 73	184 W 74	186 Re 75	190 Os 76	192 Ir 77	195 Pt 78	197 Au 79	201 Hg 80	204 Tl 81	207 Pb 82	209 Bi 83	(209) Po 84	(210) At 85	(222) Rn 86
(223) Fr 87	(226) Ra 88	(227) Ac 89															

Figure 13.2 Periodic Table showing the relative atomic mass and atomic number.

The relative atomic mass is usually equal to the number of protons + the number of neutrons in the nucleus. Some elements have a variable number of neutrons, however, and the A_r is a sort of 'average'. The most notable cases are copper (which has an A_r of 63.5) and chlorine (which has an A_r of 35.5).

If we know the relative atomic masses of the elements, then we can work out the **relative molecular masses** (M_r) of compounds. In the examples below, we get the relative atomic masses of the elements from the Periodic Table.

- **Water** (H_2O): In this molecule, there are two hydrogen atoms and one oxygen atom. The relative molecular mass is $[(2 \times 1) + 16] = 18$.
- **Carbon dioxide** (CO_2): In this molecule, there are two oxygen atoms and one carbon atom. The relative molecular mass is $[(2 \times 16) + 12] = 44$.

For ionic compounds, it is more correct to use the term **relative formula mass**, as there are no separate molecules in ionic compounds.

- **Magnesium oxide** (MgO): In this compound, there is one magnesium ion for every one oxygen ion. The relative formula mass equals $[24 + 16] = 40$.
- **Sodium carbonate** (Na_2CO_3): In this compound, there are two sodium ions, and one carbonate ion made up of one carbon and three oxygen atoms. The relative formula mass equals $[(2 \times 23) + 12 + (3 \times 16)] = [46 + 12 + 48] = 106$.

Calculating percentage composition of compounds

REVISED

We can use relative atomic masses to calculate the percentage (by mass) of different elements in a compound.

The relative atomic mass of carbon is 12 and that of oxygen is 16. The relative molecular mass of carbon dioxide (CO_2) is 44. The percentage composition (by mass) of carbon and oxygen in carbon dioxide is calculated by:

$$\text{Percentage of carbon in CO}_2 = \frac{\text{total relative mass of \textbf{carbon} in the molecule}}{\text{relative molecular mass of carbon dioxide}} \times 100\%$$

$$= \frac{12}{44} \times 100 = 27.3\%$$

$$\text{Percentage of oxygen in CO}_2 = \frac{\text{total relative mass of \textbf{oxygen} in the molecule}}{\text{relative molecular mass of carbon dioxide}} \times 100\%$$

$$= \frac{32}{44} \times 100 = 72.7\%$$

Example

Calculate the percentage composition by mass of the different elements in magnesium carbonate, $MgCO_3$.

Relative atomic masses: Mg = 24; C = 12; O = 16.

Answer

Relative formula mass of $MgCO_3$ = 24 + 12 + [3 × 16] = 84

% composition of Mg = $\frac{24}{84}$ × 100 = 28.6%

% composition of C = $\frac{12}{84}$ × 100 = 14.3%

% composition of O = $\frac{48}{84}$ × 100 = 57.1%

Mixtures and separation

Mixtures

In a mixture, the atoms or molecules of the different substances are not chemically joined. They can be separated relatively easily by physical processes such as filtration, evaporation, chromatography, and distillation.

- **Filtration** involves separating the components of a mixture according to their size, with only the smaller particles going through the filter, which retains the larger particles.
- **Evaporation** separates mixtures of solids and liquids by evaporating the liquid. The mixture is heated so that the liquid vaporises and evaporates, leaving the solid behind.
- **Chromatography** can separate substances according to their solubility in a particular solvent. It is dealt with in more detail below.
- **Distillation** is used to separate a mixture of liquids. If they have significantly different boiling points, when one component reaches its boiling point, it will evaporate at a more rapid rate (and can be condensed) leaving the remaining component(s) in liquid form. Distillation is covered later in the book.

Chromatography

Chromatography is a technique that can be used to identify and separate substances in a mixed solution.

1 A drop of the mixture is placed on chromatography paper and the paper is then placed in a container of solvent, with the level of the solvent just below the level of the spot (see Figure 13.3).
2 The solvent moves up the paper. Any soluble substance in the pigment dissolves in the solvent and travels up the paper with it.
3 The more soluble a substance is in the solvent, the further it will travel.
4 As a result, the different substances are separated on the paper. The substances can be identified by how far they have travelled. To measure this, scientists calculate something called an R_f value. Figure 13.4 shows how this is calculated.

The R_f value of a substance is different for different solvents. It is possible for two different substances to have the same R_f value in a certain solvent, but they would not have identical R_f values in several different solvents, so sometimes chromatography has to be done several times, in different solvents, to conclusively identify a solute.

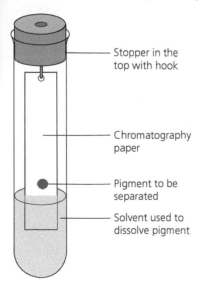

Figure 13.3 Paper chromatography.

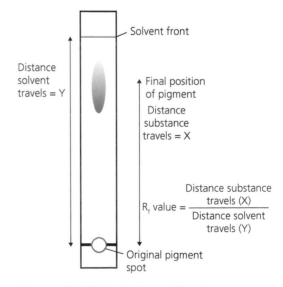

$$R_f \text{ value} = \frac{\text{Distance substance travels (X)}}{\text{Distance solvent travels (Y)}}$$

Figure 13.4 Calculation of R_f value.

Chemical reaction basics

A chemical reaction occurs when atoms in the **reactant(s)** rearrange to form one or more **products**. There will always be the same number of each type of atom in the products and the reactants. There are usually clues that a chemical reaction has taken place. For example:

● A change in colour.
● A temperature change (**exothermic** reactions give out heat, **endothermic** reactions absorb heat – this is covered later in the book)
● **Effervescence** (fizzing) when a gas is formed in a liquid.

There are two ways of representing a chemical reaction, seen in the example below – a word equation and a chemical equation.

$$CuO \quad + \quad H_2SO_4 \quad \rightarrow \quad CuSO_4 \quad + \quad H_2O$$

copper oxide   sulfuric acid   copper sulfate   water

The chemical equation is more useful because it gives us information about the atoms involved.

Chemical equations

When a chemical reaction occurs, the atoms rearrange to form products. Therefore a chemical equation must have the same number of atoms of each type on both sides of the equation, as atoms do not appear from nowhere, nor do they disappear. The total relative mass on both sides must be equal. Sometimes, just writing down the chemical symbols is all that's needed. For example:

$$H_2SO_4 \quad + \quad MgO \quad \rightarrow \quad MgSO_4 \quad + \quad H_2O$$

sulfuric acid   magnesium oxide   magnesium sulfate   water

If you count the number of the hydrogen, sulfur, oxygen and magnesium atoms on each side of the equation, they are equal. Nothing more needs to be done.

Now look at this equation:

$$H_2SO_4 \quad + \quad NaOH \quad \rightarrow \quad Na_2SO_4 \quad + \quad H_2O$$

sulfuric acid   sodium hydroxide   sodium sulfate   water

If we count the atoms on each side, we see that they do not match. On the left (the reactants), there are three hydrogen atoms in total, one sulfur, five oxygen and one sodium atom. On the right-hand side (the products) there are two hydrogen, one sulfur, five oxygen and two sodium atoms. The equation is not **balanced**, therefore it is not correct.

The equation needs to be balanced. The best way to approach this is to balance one type of atom at a time.

Let's start with the sodium. We need to have two sodium atoms on the left-hand side to balance the two on the right-hand side, so let's put a 2 before the NaOH.

$$H_2SO_4 \quad + \quad 2NaOH \quad \rightarrow \quad Na_2SO_4 \quad + \quad H_2O$$

sulfuric acid   sodium hydroxide   sodium sulfate   water

Now we need to count the atoms again (see Table 13.7).

Table 13.7 Balancing chemical equations (part 1).

Atom	Reactants	Products
Hydrogen	4	2
Sulfur	1	1
Oxygen	6	5
Sodium	2	2

Now the sodium atoms balance, but the hydrogen still doesn't, and the balance of the oxygens has been thrown out. Now let's address the hydrogens. We can balance them by putting a 2 in front of the H_2O.

$$H_2SO_4 \quad + \quad 2NaOH \quad \rightarrow \quad Na_2SO_4 \quad + \quad 2H_2O$$

sulfuric acid   sodium hydroxide   sodium sulfate   water

Now let's count the atoms again (Table 13.8).

Table 13.8 Balancing chemical equations (part 2).

Atom	Reactants	Products
Hydrogen	4	4
Sulfur	1	1
Oxygen	6	6
Sodium	2	2

You will see that by doing that, not only have we balanced the hydrogens, but the oxygens are now balanced as well. In fact, the whole equation is balanced.

Note that in chemical equations the state of each component is often indicated in brackets after the formula, using the following code:
- (s) = solid
- (l) = liquid
- (g) = gas
- (aq) = aqueous solution (i.e. dissolved in water)

Numbers in chemical equations

People sometimes get confused about the numbers before a symbol in a chemical equation and the subscript numbers. Look at the water in the balanced equation above ($2H_2O$). The 2 in front of the formula indicates how many molecules of water are involved in the reaction, and that number can be altered to balance the equation. The 2 after the H indicates that, in a water molecule, there are two atoms of hydrogen. That can never be altered to balance an equation, because you would end up with a chemical formula that is either for a different chemical, or probably does not exist at all.

Now test yourself

TESTED

6 State three observations that can indicate that a chemical reaction is taking place.
7 Balance the following chemical equations.
   a) $Cl_2 + NaBr \rightarrow Br_2 + NaCl$
   b) $MgCO_3 \rightarrow MgO + CO_2$
   c) $CH_4 + O_2 \rightarrow CO_2 + H_2O$
   d) $CaCl_2 + NaOH \rightarrow Ca(OH)_2 + NaCl$
8 What does (aq) after a chemical symbol in an equation mean?

Answers online

Chemical calculations

Calculating the formula for a compound

We can use data from chemical reactions to calculate the formula of a compound. What is calculated is the simplest form of the formula, known as the **simple formula**. Let's look at some examples.

We want to find the formula of copper oxide. 2.5 g of copper is heated to form copper oxide. The mass of copper oxide produced is 3.13 g.

1. We know the mass of copper (2.5 g). That copper will be part of the copper oxide, and the only other element present is oxygen. So, the mass of oxygen in the copper oxide is $(3.13 - 2.5) = 0.63$ g.
2. Now we look up the A_r of copper and oxygen in the Periodic Table. Copper = 63.5, oxygen = 16.
3. Now divide the masses by the A_r. For copper, $\frac{2.5}{63.5} = 0.04$. For oxygen, $\frac{0.63}{16} = 0.04$.
4. Now divide the answers by the smallest number (0.04). For copper, $\frac{0.04}{0.04} = 1$. For oxygen, $\frac{0.04}{0.04} = 1$. This gives us the ratio of copper to oxygen in the formula.
5. The ratio is $1:1$, so the formula is CuO.

We want to find the formula of sodium sulfide. 1.15 g of sodium reacts with 0.8 g of sulfur to form sodium sulfide.

1. We know the mass of sodium (1.15 g) and sulfur (0.8 g).
2. Now we look up the A_r of sodium and sulfur in the Periodic Table. Sodium = 23, sulfur = 32.
3. Now divide the masses by the A_r. For sodium, $\frac{1.15}{23} = 0.05$. For sulfur, $\frac{0.8}{32} = 0.025$.
4. Now divide the answers by the smallest number (0.025). For sodium, $\frac{0.05}{0.025} = 2$. For sulfur, $\frac{0.025}{0.025} = 1$. This gives us the ratio of sodium to sulfur in the formula.
5. The ratio is $2:1$, so the formula is Na_2S.

In these examples, step 3 is calculating the number of **moles** of each substance. Moles are covered later in the book.

> **Exam tip**
>
> You cannot learn how to do chemical calculations just by reading about them. You must work through examples. It does not matter if you do the same example several times, as you still learn the process.

Calculating masses of reactants and products

Chemical equations also allow us to calculate the masses of the products (if we know the masses of the reactants) or the masses of the reactants needed to make a given mass of product. An example is the reaction of magnesium with oxygen.

$$2Mg(s) \quad + \quad O_2(g) \quad \rightarrow \quad 2MgO$$

magnesium   oxygen   magnesium oxide

We want to predict how much magnesium oxide you would get in this reaction if you used 5 g of magnesium.

- The relative atomic mass of magnesium is 24 and oxygen is 16.
- This means that the relative formula mass of magnesium oxide is 40.
- The equation tells us that (2×24) g of magnesium produces (2×40) g of magnesium oxide.
- 48 g of magnesium produces 80 g of magnesium oxide.

H ● 1 g of magnesium will produce $\frac{80}{48}$ g (=1.7 g) of magnesium oxide.

● 5 g of magnesium will produce $5 \times 1.7 = 8.5$ g of magnesium oxide.

8.5 g is the **theoretical yield** of magnesium oxide, but chemical reactions are never 100% efficient so the actual yield will be lower than this. The actual yield cannot be calculated, it must be measured.

We can also reverse the process to calculate the mass of reactants that would be needed to form a given mass of product. In the reaction above, suppose we wanted to know how much magnesium would be needed to give 10 g of magnesium oxide.

● The relative atomic mass of magnesium is 24 and oxygen is 16.
● This means that the relative formula mass of magnesium oxide is 40.
● The equation tells us that (2×24) g of magnesium produces (2×40) g of magnesium oxide.
● 48 g of magnesium produces 80 g of magnesium oxide.
● 1 g of magnesium oxide requires $\frac{48}{80}$ g (=0.6 g) of magnesium.
● To get 10 g of magnesium oxide, you need $10 \times 0.6 = 6$ g of magnesium.

Once again, this calculation assumes that the reaction is 100% efficient. In practice, slightly more than 6 g of magnesium would be needed to make 10 g of magnesium oxide.

Yields and moles

Calculating the yield of a chemical reaction

In any chemical manufacturing process, it is useful for scientists to know what yields are possible, as they can then check how efficient their process is and to what extent further improvements can be made. It is therefore useful to work out the **theoretical yield** of a reaction, and what percentage of that theoretical target is being achieved – the **percentage yield**.

We will use the example given in the last section, for the burning of magnesium in air to give magnesium oxide.

We calculated that 5 g of magnesium gave a theoretical yield of 8.5 g of magnesium oxide. Suppose when this was tested, only 7.9 g of magnesium oxide were produced. 7.9 g is the actual yield.

The percentage yield is calculated as follows:

$$\text{Percentage yield} = \frac{\text{actual yield}}{\text{theoretical yield}} \times 100$$

$$\text{Percentage yield} = \frac{7.9}{8.5} \times 100 = 93\%$$

Note that actual yield can be measured in various ways (grams, moles, tonnes, etc.). The actual and theoretical yield must have the same units when calculating percentage yield. This is rarely a problem as the masses of reactants and products are nearly always expressed in the same units, which are then used to calculate the theoretical yield.

The Avogadro constant and the mole

REVISED ☐

The Avogadro constant is a number and it is 6.02×10^{23}. It is named after an Italian chemist, Lorenzo Romano Amedeo Carlo Avogadro. It is the number of carbon-12 (^{12}C) atoms in exactly 12 g of carbon-12. ^{12}C is used to establish the relative atomic masses of all the other elements. We cannot weigh atoms, but we know that a carbon atom has 6 protons and 6 neutrons, and that protons weigh the same as neutrons. So, we can say that each proton and each neutron weighs '1'. This means that the relative atomic mass (of elements that exist as single atoms) or the relative molecular mass of molecules, in grams, contain a number of atoms/molecules equal to Avogadro's constant. This leads to the concept of the mole, which allows us to relate numbers of atoms and molecules to grams.

One mole is the amount of pure substance containing the same number of chemical units as there are atoms in exactly 12 g of carbon-12 (i.e. 6.02×10^{23}, Avogadro's constant). For atoms, it is equal to the relative atomic mass in grams. For compounds or molecules, it is equal to the relative molecular mass in grams.

Now test yourself

TESTED ☐

9 Mercury chloride can be converted into mercury. 5.44 g of mercury chloride produce 4.02 g of mercury. What is the formula of mercury chloride?

10 When heated, potassium chlorate decomposes into potassium chloride and oxygen.
$2KClO_3 \rightarrow 2KCl + 3O_2$
What will be the theoretical yield of potassium chloride if 10 g of potassium chlorate was used? (Relative atomic masses: K = 39; Cl = 35.5; O = 16.)

11 Nitrogen and hydrogen are reacted together to produce ammonia.
$N_2 + 3H_2 \rightarrow 2NH_3$
60 g of hydrogen is reacted with an excess of nitrogen. What is the theoretical yield of ammonia? (Relative atomic masses: N: 14; H = 1.)

12 In the reaction described in question 3, 125 g of ammonia was produced. What is the percentage yield of this reaction?

13 How many moles of calcium carbonate ($CaCO_3$) are there in 50 g? (Relative atomic masses: Ca = 40; C = 12; O = 16.)

Answers online

Summary

- Elements are substances that cannot be broken down into simpler substances by chemical means and are the basic building blocks of all substances.
- Elements are made up of only one type of atom.
- Compounds are substances made of two or more different types of atom that are chemically joined and have completely different properties to their constituent elements.
- Elements can be represented using chemical symbols and simple molecules using chemical formulae.
- The formulae of ionic compounds can be established given the formulae of the ions they contain.
- Relative atomic mass and relative molecular (formula) mass are used to compare the masses of different elements and different compounds.
- The percentage composition of compounds can be calculated from the chemical formula.
- Atoms/molecules in mixtures are not chemically joined and mixtures are easily separated by physical processes such as filtration, evaporation, chromatography and distillation.

- The R_f values of components of mixtures can be used to identify them.
- Chemical reactions involve the re-arrangement of the atoms present in the reactants to form one or more products, which have the same total number of each type of atom as the reactants.
- Colour changes, temperature changes (exothermic/ endothermic) and effervescence can be used as evidence that a chemical reaction has taken place.
- Chemical reactions can be represented using word equations.
- Chemical reactions can be represented using balanced chemical equations, where the total relative mass of reactants and products is equal.
- (H) The formula of a compound can be calculated from reacting mass data.
- The masses of reactants or products can be calculated from a balanced chemical equation.
- The percentage yield of a chemical reaction can be calculated from a measured actual yield.
- The Avogadro constant represents the number of atoms in 12 g of carbon-12.

Exam practice

(H) 1 Magnesium is burnt in a crucible and reacts with oxygen to form magnesium oxide. The experimenter ensures that all the magnesium has reacted. The crucible is weighed before the experiment, after adding the magnesium, and at the end of the experiment. The results are shown below.
Mass of crucible: 35.6 g
Mass of crucible + magnesium: 41.9 g
Mass of crucible and magnesium oxide: 46.0 g
Relative atomic masses: Mg = 24, O = 16
   a) Use the data above to calculate the formula of magnesium oxide. Show your working. [5]
   b) Explain why this calculation would be inaccurate if the magnesium had not all reacted. [2]
   c) Calculate the percentage of magnesium in magnesium oxide. [2]
2 Nitrogen reacts with hydrogen to form ammonia. The equation is shown below, but it is not balanced.

$$N_2 + H_2 \rightarrow NH_3$$

   a) Balance the equation. [2]
   b) Calculate the theoretical yield of ammonia (to the nearest whole number of grams) if 100 g of nitrogen was used in this reaction. [2]
   (Relative atomic masses: N = 14; H =1.)
   c) When the reaction was carried out, 26.6 g of ammonia was produced. Calculate the percentage yield of the reaction. [2]

Answers and quick quiz 13 online

ONLINE

14 Atomic structure and the Periodic Table

Atoms and isotopes

Atomic structure

Elements are made up of atoms, and all the atoms of an element are of the same type. You cannot understand chemical properties and reactions properly unless you know about the structure of atoms, and about the smaller particles that make up an atom.

Every atom is made up of smaller particles called **protons**, **neutrons** and **electrons**. These are referred to as fundamental particles, and their numbers vary in atoms of different elements.

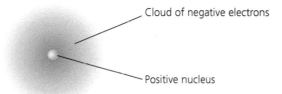

Cloud of negative electrons

Positive nucleus

Figure 14.1 The structure of an atom.

The structure of an atom is shown in Figure 14.1. Every atom contains a positively charged central region called the **nucleus**. This is surrounded by light, negatively charged electrons. The positive and negative charges balance out so that the atom is electrically neutral and has no overall charge. Here are some important facts about atoms.

- The nucleus makes up nearly all the mass of the atom, and has two types of particle within it: protons, which have a positive charge, and neutrons, which have no charge.
- The mass of a neutron is the same as that of the proton.
- Each electron has a negative charge.
- The mass of an electron is so small that it is considered to be negligible.
- We call the mass of a proton 1 unit and its charge +1, and describe the other particles relative to these values.

There are two other terms related to the structure of atoms which you need to know:

- **Atomic number:** the number of protons in the nucleus. The number of electrons in an atom is always equal to the number of protons.
- **Mass number:** the **total** number of nucleons (protons + neutrons) in the nucleus.

Sometimes, the chemical symbol for an element is written in a way that shows the atomic number and the mass number. An example is given in Figure 14.2, for the element sodium.

Mass number ——— $^{23}_{11}$**Na**

Atomic number ———

Figure 14.2 Chemical symbol for sodium.

> **Exam tip**
>
> There can be confusion about the mass number and the relative atomic mass. The mass number is the number of protons and neutrons in a specific atom of the element. The relative atomic mass is the average mass number of all the isotopes of an element. Most Periodic Tables give the relative atomic mass, not the mass number.

Isotopes

Different elements have different atomic (proton) numbers, and no two elements have the same number of protons. For any given element, the atomic number is always the same. However, the number of neutrons in the nucleus is not fixed. Some atoms of the same element have different numbers of neutrons and so they have different mass numbers. These different forms of the same element are called **isotopes**.

For example, most carbon atoms have 6 protons and 6 neutrons (mass number 12), but another form of carbon has 8 neutrons (mass number 14). The two isotopes are referred to as carbon-12 and carbon-14, and represented by symbols as ^{12}C and ^{14}C.

When isotopes have more neutrons than protons, the atom is sometimes unstable and likely to decay. Such isotopes are radioactive. The more the number of neutrons differs from the stable form, the more likely it is that the atom will be radioactive. In the example of carbon given above, ^{12}C is stable but ^{14}C is unstable and radioactive.

The relative atomic mass of an element (A_r) is the average of the mass numbers of all its isotopes, taking into account their frequency. For example, the A_r of chlorine is 35.5. That is because there are two isotopes, ^{35}Cl and ^{37}Cl. The relative atomic mass is 35.5 and not 36 because 75% of chlorine atoms are ^{35}C and 25% are ^{37}Cl.

> **Exam tip**
>
> You will not be tested on this paragraph.

Electrons and electron shells

The electrons in an atom are in different **electron shells** around the nucleus. The shells are sometimes called **orbits**. Each shell can contain only a certain number of electrons as shown in Table 14.1.

Table 14.1 The number of electrons held by different shells.

Shell (or orbit)	Maximum number of electrons accommodated for elements hydrogen to calcium
1	2
2	8
3	8
4	2

Electron structure – worked example

The element sodium, Na, has an atomic number of 11. This means that it has 11 protons in the nucleus, and so must have 11 electrons surrounding the nucleus. The first shell will be full (2 electrons), the second will be full (8 electrons) and the one remaining electron will be in the third shell. The electronic structure for sodium can be written as 2,8,1 and it can be depicted as in Figure 14.3.

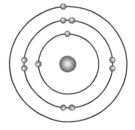

Figure 14.3 Electronic structure of a sodium atom.

Answers and quick quizzes at www.hoddereducation.co.uk/myrevisionnotesdownloads

Electronic structure and the Periodic Table

Electronic structure

REVISED

Table 14.2 shows the electronic structure for the first 20 elements in the Periodic Table. Note that each time the atomic number goes up, this means there is an extra proton. As the number of electrons always equals the number of protons, an extra electron is also added. This goes into the next available position in a shell or, if the shell is full, into the first position of a new shell.

Table 14.2 Electronic structures of the first 20 elements of the Periodic Table.

Atomic number	Element	Shell				Atomic number	Element	Shell			
		1	2	3	4			1	2	3	4
1	Hydrogen	1				11	Sodium	2	8	1	
2	Helium	2				12	Magnesium	2	8	2	
3	Lithium	2	1			13	Aluminium	2	8	3	
4	Beryllium	2	2			14	Silicon	2	8	4	
5	Boron	2	3			15	Phosphorus	2	8	5	
6	Carbon	2	4			16	Sulfur	2	8	6	
7	Nitrogen	2	5			17	Chlorine	2	8	7	
8	Oxygen	2	6			18	Argon	2	8	8	
9	Fluorine	2	7			19	Potassium	2	8	8	1
10	Neon	2	8			20	Calcium	2	8	8	2

Now test yourself

TESTED

1 What does the atomic number of an element signify?
2 An atom of phosphorus has an atomic number of 15 and a mass number of 31. How many neutrons and how many electrons are there in the atom?
3 What is an isotope?
4 The element boron has 5 protons. There are 14 isotopes of boron, two of which are ^{11}B and ^{16}B. Suggest which of these two isotopes is the most stable, giving a reason for your answer.
5 The atomic number of silicon is 14. What is its electronic structure?

Answers online

The Periodic Table

REVISED

The Periodic Table is a chart showing all the known elements arranged in a logical way that allows chemists to predict the properties of individual elements. The elements are arranged in order of their atomic numbers. The Periodic Table is shown in Figure 14.4.
- Each column is called a **group**.
- Each horizontal row of elements is called a **period**.
- The elements in any one group have similar properties.

Hydrogen has some unusual properties and so is not included in a group. Helium is placed with hydrogen in the first period which only contains two elements. In between Groups 2 and 3 are a collection of metals known as the **transition metals**. At the junction between metals and non-metals in the table are certain elements which have properties that are intermediate between those of metals and non-metals (e.g. silicon).

Group 1 (I)	Group 2 (II)											Group 3 (III)	Group 4 (IV)	Group 5 (V)	Group 6 (VI)	Group 7 (VII)	Group 0
						1 H 1											4 He 2
7 Li 3	9 Be 4											11 B 5	12 C 6	14 N 7	16 O 8	19 F 9	20 Ne 10
23 Na 11	24 Mg 12											27 Al 13	28 Si 14	31 P 15	32 S 16	35.5 Cl 17	40 Ar 18
39 K 19	40 Ca 20	45 Sc 21	48 Ti 22	51 V 23	52 Cr 24	55 Mn 25	56 Fe 26	59 Co 27	59 Ni 28	63.5 Cu 29	65 Zn 30	70 Ga 31	73 Ge 32	75 As 33	79 Se 34	80 Br 35	84 Kr 36
85 Rb 37	88 Sr 38	89 Y 39	91 Zr 40	93 Nb 41	96 Mo 42	98 Tc 43	101 Ru 44	103 Rh 45	106 Pd 46	108 Ag 47	112 Cd 48	115 In 49	119 Sn 50	122 Sb 51	128 Te 52	127 I 53	131 Xe 54
133 Cs 55	137 Ba 56	139 La 57	178 Hf 72	181 Ta 73	184 W 74	186 Re 75	190 Os 76	192 Ir 77	195 Pt 78	197 Au 79	201 Hg 80	204 Tl 81	207 Pb 82	209 Bi 83	(209) Po 84	(210) At 85	(222) Rn 86
(223) Fr 87	(226) Ra 88	(227) Ac 89															

Figure 14.4 The modern Periodic Table, showing symbols, atomic numbers and relative atomic masses (elements 58 to 71, the lanthanides or rare earth elements, and elements with atomic number greater than 89 have been omitted for simplicity).

Electronic structure and the Periodic Table

REVISED

The Periodic Table arranges elements in order of their atomic number.

- The atomic number indicates the number of protons in the nucleus but, as the numbers of protons and electrons in an element are the same, the elements are effectively arranged in order of their number of electrons too.
- The group number indicates the number of electrons in the outer shell of the atom. Group 0 is odd in this respect. In that group, the atoms have full outer shells. The problem is that the number varies. Most of Group 0 have eight electrons in their outer shell, but helium only has two (because the first shell can only contain two electrons). So, in effect, Group 0 indicates that there are no electrons in the 'outer' shell, and the one below it is full.
- The period indicates the number of occupied electron shells in an atom.

This information gives you the electronic structure of any element. For example, aluminium has three occupied electron shells, because it is in period 3, and three electrons in its outer shell because it is in Group 3. The first two shells are full; the first containing 2 electrons and the second containing 8. The electronic structure of aluminium is as shown in Figure 14.5. The fact that elements in a group have the same number of electrons in their outer shell gives them similar properties, and the increasing number of shells as you go down the group creates trends in those properties.

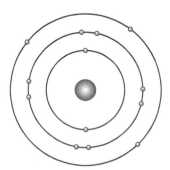

Figure 14.5 Electronic structure of aluminium.

> **Exam tip**
>
> You may wonder what is going on with the transition metals, between Groups 2 and 3. Their electronic structure is complex and not covered at GCSE, so you will not be asked about them.

Answers and quick quizzes at www.hoddereducation.co.uk/myrevisionnotesdownloads

Group 1 and Group 7

The elements of Group 1

There are six elements in Group 1 of the Periodic Table; these are the **alkali metals** (so called because when they react with water they form alkaline solutions).

- **Lithium** is a soft, silvery metal. It is the lightest of all metals and the least dense solid element. Although it is the least reactive alkali metal, it is still sufficiently reactive that it has to be stored in mineral oil.
- **Sodium** is a soft, silvery-white metal. It is not found freely in nature because it is so reactive. For this reason, it is stored under mineral oil to avoid it reacting with air.
- **Potassium** is a soft, silvery-grey metal. The alkali metals get more and more reactive as you go down the group, so potassium also has to be stored under oil, but even then it can sometimes react with the small amount of oxygen that dissolves in oil.
- **Rubidium** is another soft, silvery and highly reactive metal. If it gets into contact with air it will burst into flames, so once again it is always stored under oil.
- **Caesium** is so reactive that not only does it have to be kept in oil, but it can only be handled in an inert atmosphere. If it comes into contact with water, it will explode violently.
- **Francium** is the most reactive element in this highly reactive group. It is extremely radioactive and unstable. Although found naturally in uranium minerals, there is only about 30 g in existence on the Earth at any one time.

Properties of the Group 1 elements

Elements in the same group have similar properties. The properties of the alkali metals are as follows.
- They react with oxygen to form oxides.
- They react with water to form hydroxides and hydrogen gas.
- They react with chlorine and bromine to form chlorides and bromides.

The reactivity of the alkali metals increases down the group, so the reactions listed get more vigorous (sometimes violently so) as you move from lithium to potassium. The other alkali metals are so reactive that these reactions would be highly dangerous to carry out.

Group 1 reaction equations

In these equations, lithium is used as an example. For the other members of Group 1, their formula can be substituted for Li.

Reaction with oxygen:

$$4Li(s) \quad + \quad O_2(g) \quad \rightarrow \quad 2Li_2O(s)$$
lithium $\qquad\qquad$ oxygen $\qquad\qquad$ lithium oxide

Reaction with water:

$$2Li(s) \quad + \quad 2H_2O(l) \quad \rightarrow \quad 2LiOH(aq) \quad + \quad H_2$$
lithium $\qquad$ water $\qquad$ lithium hydroxide $\qquad$ hydrogen

The hydrogen given off can be identified by testing with a lighted splint. It burns with a 'squeaky pop'.

Reactions with halogens:

$$2Li(s) \quad + \quad Cl_2(g) \quad \rightarrow \quad 2LiCl(s)$$
lithium   chlorine   lithium chloride

$$2Li(s) \quad + \quad Br_2(g) \quad \rightarrow \quad 2LiBr(s)$$
lithium   bromine   lithium bromide

Flame tests for alkali metals

The three alkali metals burn in oxygen with different coloured flames.
- Lithium burns crimson-red.
- Sodium burns orange-yellow.
- Potassium burns lilac.

All the salts of the alkali metals burn with the same colour flames as the metals, and so the flame test can be used as a test for whether a compound is an alkali metal salt. The usual way of carrying out a flame test is to dip a metal probe into $2\,mol\,dm^{-3}$ hydrochloric acid and then place the probe into a roaring Bunsen flame. The hydrochloric acid is used to convert the salts into chlorides, which are particularly volatile and so work well in flame tests.

> **Volatile:** Evaporates easily at normal temperatures.

Flame tests can also be used for other metal ions. Calcium (Ca^{2+}) burns with a brick-red flame and barium (Ba^{2+}) produces an apple-green flame.

Now test yourself

TESTED

6 What information about an atom of an element can be obtained by the period it is in in the Periodic Table?
7 Sulfur is in Group 6 and period 3 of the Periodic Table. How many electrons it has got?
8 Draw the electronic structure of magnesium (Mg) using information from the Periodic Table (Figure 2.4).
9 Put these alkali metals in order of reactivity, starting with the most reactive: potassium, lithium, rubidium.
10 What substances are formed when potassium reacts with water?

Answers online

The Group 7 elements

REVISED

The Group 7 elements are the non-metals known as the **halogens**.
- **Fluorine** is a pale yellow gas at room temperature. It is highly toxic (air with just 0.1% of fluorine in it can be lethal) and explosive. Despite that, very small quantities of its salts (fluorides) are essential in the human body.
- **Chlorine** is a greenish-yellow gas. It is very reactive and toxic, but because of its reactivity it is most commonly found in its salts (chlorides), including sodium chloride that we know as salt. It is used in bleach, in the manufacture of the plastic PVC, and to disinfect drinking water and the water in swimming pools.
- **Bromine** is a reddish-brown liquid at room temperature. Once again, it is both reactive and toxic.
- **Iodine** is less reactive than the halogens we have seen so far, though it is still a reactive element and for this reason is most usually found in its salts (iodides). It is toxic in its elemental form, but it is essential in the human body in small quantities for the proper working of the thyroid gland. It is used in antiseptics, disinfectants, dyes and printing inks.

- **Astatine** is one of the rarest elements on Earth. It is highly radioactive and breaks down so rapidly that it cannot be described – because no-one has ever seen it!

Just like the Group 1 elements, the halogens show similar properties to one another, and there is a trend of increasing reactivity in the group. In this case, however, reactivity increases as you go up the group, not down as with the alkali metals.

Reactivity of halogens

Reactions of the halogens

The reactions of chlorine with lithium and sodium were covered in the section on alkali metals. The other halogens react with the alkali metals in similar ways, forming their respective salts.

Reactions with iron

Fluorine is a very reactive gas. Iron wool bursts into flames without heating when fluorine gas flows over it. The reaction produces a halide called iron(III) fluoride.

$$2Fe(s) \quad + \quad 3F_2 \quad \rightarrow \quad 2FeF_3$$

iron $\qquad$ fluorine $\qquad$ iron(III) fluoride

The reactivity of the halogens can be illustrated by their reactions with iron. Iron is heated with each halogen. The observations are as follows:

- Chlorine: The iron glows brightly and a blackish-brown deposit of iron(III) chloride is formed in the tube. See Figure 14.6.
- Bromine: The apparatus is set up as shown in Figure 14.7. The iron glows slightly and a brown residue is left (iron(III) bromide).
- Iodine: If the bromine experiment is repeated with iodine, strong heating is required before any reaction occurs.

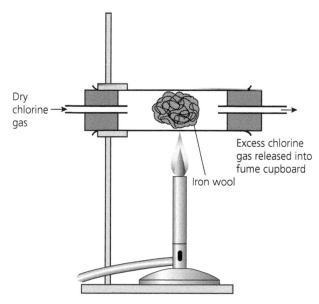

Figure 14.6 Set-up to burn iron in chlorine.

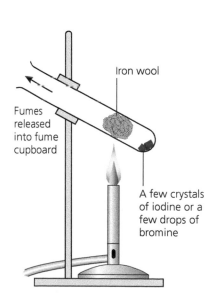

Figure 14.7 Set-up for combustion of bromine/iodine with iron.

Displacement reactions

A more reactive halogen can displace a less reactive halogen from a solution of its salts. This is referred to as a **displacement reaction**.

Chlorine, the second most reactive halogen (after fluorine), will displace both bromine and iodine from solutions of bromides and iodides.

$$Cl_2(g) \quad + \quad 2NaBr(aq) \quad \rightarrow \quad Br_2(l) \quad + \quad 2NaCl(aq)$$

chlorine   sodium bromide   bromine   sodium chloride

The bromine has been displaced from sodium bromide and replaced by chlorine to form sodium chloride. In the same way, bromine, which is more reactive than iodine, displaces iodine from iodide solutions. In summary:
- Chlorine displaces bromine and iodine from their salts.
- Bromine displaces iodine but is displaced by chlorine.
- Iodine does not displace any of the other halogens and is displaced by bromine and chlorine.

Tests for halogens

Halides are identified by a precipitation reaction with **silver nitrate**. Chloride, bromide and iodide ions combine with silver ions to form insoluble precipitates of white silver chloride, pale yellow silver bromide and yellow silver iodide, respectively. The different colours of the precipitates mean that the halides can be identified.

Silver nitrate and halides are ionic compounds, and form ions in solution. The silver ions react with the halide ions to form a silver salt, for example with chlorine.

$$Ag^+(aq) + Cl^-(aq) \rightarrow AgCl(s)$$

The silver chloride is solid and so forms a precipitate.

The test is carried out as follows:
1 Add dilute nitric acid to the test solution (volume equal to the volume of test solution).
2 Add a few drops of silver nitrate solution. Observe the colour of the precipitate.
   - White – salt was a chloride.
   - Pale yellow – the salt was a bromide.
   - Yellow – the salt was an iodide.

The nitric acid is added to remove other ions, such as carbonate or hydroxide, which could interfere with the reaction.

The colours of the three precipitates are quite similar, and a further step is usually carried out to confirm the result. This involves trying to dissolve the precipitate in **ammonia solution**.
- Chlorides dissolve in dilute ammonia solution.
- Bromides are insoluble in dilute ammonia but soluble in concentrated ammonia solution.
- Iodides are insoluble in dilute and concentrated ammonia.

> **Precipitate:** An insoluble solid that is formed in a liquid. The appearance of the insoluble solid from solution is called precipitation.

The reactivity of alkali metals and halogens explained

The different reactivities of the alkali metals and the halogens is a result of their electronic structure.

Many chemical reactions involve atoms losing or gaining electrons to form positive or negative **ions**.

Group 1 elements have one electron in their outer shell, and Group 7 elements have seven electrons. An atom will be most stable when its outer shell of electrons is full (i.e. eight electrons in all shells apart from the first, which can only contain two). If Group 1 elements lose their outer shell electron, then their new outer shell (i.e. the one below) is full. If Group 7 elements gain one electron, then their outer shell is full. It is very easy to lose or gain just one electron, so these elements will react very readily.

Group 2 and Group 6 elements have to lose or gain two electrons to achieve a full outer shell. This is more difficult. Once one electron has been lost, it would give the atom an overall positive charge, which will make it more difficult for a second electron to be lost, as electrons are negative and are attracted to positive charges. Exactly the reverse is the case if two electrons need to be gained – gaining one electron creates an overall negative charge, which will tend to repel any further electrons. For this reason, the elements in Groups 2 and 6 are less reactive than those in Groups 1 and 7.

The different reactivities of the alkali metals and the halogens is also a result of their electronic structure. The further away an electron shell is from the positive nucleus, the easier it is for an electron to be lost and the more difficult it is to gain one.
- Alkali metals lose an electron and so the ones with more electron shells lose it more easily, i.e. reactivity increases down the group.
- The halogens gain an electron. This is easier if there are fewer shells, because the pull of the positive nucleus is greater. That is why halogen reactivity increases as you go up the group.

Group 0

Group 0 elements (sometimes called the noble gases) are very unreactive as they have a full outer shell of electrons. In nature, they do not form compounds, although scientists have been able to make some compounds of noble gases under artificial conditions.

Several Group 0 gases have commercial uses.
- **Helium** is lighter than air, and it is used in airships and balloons. Liquid helium (at −269°C) is used to cool the superconducting magnets in MRI scanners (which produce images of the body to enable doctors to identify internal problems such as tumours in patients).
- **Neon** is used in 'neon lights', although the term is also used for lighting that uses other gases. Neon produces an orange-red glow when sealed in a tube with a high potential applied to electrodes at both ends, so light of other colours are not truly neon lights.
- **Argon** is also used in lighting (it produces a blue light). As it is an even better insulator than air, it is also used in double glazing.

Now test yourself

11 Which is the least reactive halogen?
12 What substance is formed when iron is heated with chlorine?
13 What substances would be formed if you reacted chlorine with potassium iodide?
14 Which chemical is used to test for halide ions in solution?
15 Why are the Group 0 elements very unreactive?

Answers online

Summary

- Atoms have a positively charged nucleus with orbiting negatively charged electrons.
- Atomic nuclei contain protons and neutrons.
- Protons and neutrons are of equal mass (designated as 1); electrons have a negligible mass.
- Protons have a positive charge and electrons have a negative charge. Neutrons have no charge.
- Atoms have no overall electrical charge.
- The number of electrons in an atom is equal to the number of protons.
- Atomic number is the number of protons (and therefore electrons) in the element's nucleus; mass number is the number of protons + neutrons in the nucleus.
- Elements may exist in different forms with different atoms having different numbers of neutrons. These forms are called isotopes.
- The relative atomic mass of an element with more than one isotope is calculated as an average of the relative atomic masses of the isotopes, taking account of the relative frequencies of the isotopes.
- Elements are arranged in order of increasing atomic number and in groups and periods in the Periodic Table, with elements having similar properties appearing in the same groups.
- Metals are found to the left and centre of the Periodic Table and non-metals to the right, with elements having intermediate properties appearing between the metals and non-metals in each period.
- The electronic structure of any element is related to its position in the Periodic Table.
- Elements in the same group in the Periodic Table have similar physical and chemical properties (illustrated by Group 1 and Group 7 in this chapter).

- Many reactions, including those of Group 1 elements and many of those of Group 7 elements, involve the loss or gain of electrons and the formation of charged ions.
- Group 1 elements become more reactive as you go down the group, whereas Group 7 elements become more reactive as you go up the group.
- These trends in reactivity of Group 1 and Group 7 elements are related to their readiness to lose or gain an electron.
- Alkali metals react with oxygen to form oxides, with halogens to form halides, and with water to form hydroxides and hydrogen.
- The hydrogen given off when alkali metals react with water can be identified by using a lighted splint; the gas burns with a squeaky pop.
- Halogens react with iron to form iron(III) halides.
- The reactivities of chlorine, bromine and iodine can be demonstrated by displacement reactions.
- The properties of chlorine and iodine make them useful in a variety of ways.
- Alkali metal ions (and some others) can be identified by flame tests.
- Cl^-, Br^- and I^- ions can be identified by their reactions with silver nitrate solution.
- Group 0 gases have full outer electron shells and as a result are very unreactive.
- The properties of helium, neon and argon make them useful in a variety of ways.

Exam practice

1 a) The electronic structure of aluminium (Al) is 2,8,3. Using the Periodic Table, give the electronic structures of the following elements: [4]
     i)   carbon (C)
     ii) potassium (K)
     iii) magnesium (Mg)
     iv) chlorine (Cl).
  b) Phosphorus has an atomic number of 15 and a mass number of 31. In an atom of phosphorus, give the numbers of: [4]
     i)   protons
     ii) neutrons
     iii) electrons
     iv) electron shells.
  c) Use the Periodic Table to answer the following questions.
     i)   Give the chemical symbol of a non-metal found in period 2 of the Periodic Table. [1]
     ii) Give the chemical symbol of an element which you would expect to have similar properties to magnesium (Mg). [1]
     iii) Give the chemical symbol of an element that has an equal number of protons and neutrons. [1]
  d) Chlorine has a relative molecular mass of 35.5. Explain how that figure is arrived at. [4]

2 Substance X is a white crystalline solid. When placed in a roaring Bunsen flame, it burns with an orange-yellow flame. When it is dissolved in water and a few drops of silver nitrate are added, a yellow precipitate is formed.
  a) What is X? [1]
  b) What is the yellow precipitate? [1]
  c) Chlorine gas is bubbled through a solution of X. Write a balanced symbol equation for the reaction that occurs. [2]
  d) What name is given to the type of reaction that occurs in (c)? [1]

3 The gases in Group 0 are very unreactive. Two of them, helium and argon, are among the most common gases in the universe. Explain the reasons why these gases are unreactive, and so common. [6] [QER]

Answers and quick quiz 14 online

ONLINE

15 Water

Drinking water

Composition of natural water

REVISED

Water is a solvent for many substances, and natural water is never pure, nor 'clean'. Natural water supplies contain the following components:
- solutes
  - ○ ions (mainly sodium, calcium, magnesium, sulfate, hydrogencarbonate and chloride).
  - ○ dissolved gases (the most important ones are oxygen and carbon dioxide).
- microorganisms
- pollutants (e.g. fertilisers and pesticides from farm land, various industrial pollutants).

Water consumption

REVISED

Over the last 100 years, water consumption has steadily risen across the world. Some of this rise has been due to increased use of water in homes, but there has also been a steep rise in the amount of water used by industry. In the home, people are encouraged to reduce water consumption by taking the following measures:
- Having showers rather than a bath, and taking shorter showers.
- Re-using bath or sink water for watering plants.
- Having toilets with a setting to allow a shorter flush.
- Insulating water pipes to reduce the chances of bursting in winter.
- Not leaving water running when cleaning your teeth or rinsing vegetables.
- Having a water meter fitted so that water use can be monitored.
- Ensuring your washing machine or dishwasher has a full load before using it.

> **Exam tip**
>
> You need to be aware of the need to conserve water, and it is best to know a few examples of how this might be done. You do not need to learn the whole list of conservation methods given here.

Abstracting water

REVISED

Abstraction is the term used for extracting water (either temporarily or permanently) for human use. This includes things like pumping underground water to the surface, building dams and reservoirs, taking water from rivers or lakes, collecting rainwater and **desalinating** sea water.

Some of these measures may be opposed by local populations because of their environmental impact. This is especially the case with the building of dams and reservoirs, both of which require the flooding of large areas of land, destroying habitats and altering the appearance of the area.

Water distribution

REVISED

Water can only be extracted or collected in certain places. Some locations may be many miles away, and water must be piped to those areas to provide irrigation, sanitation and drinking water. This can be expensive and unfortunately some of the driest countries in the world are also some of the poorest. Approximately 1.1 billion people in the world have no access to treated drinking water, and the use of untreated water is responsible for millions of cases of death and disease every year. Water is not only needed for drinking, but also for irrigation so crops can be grown and starvation is less likely.

Water treatment

REVISED

Water for drinking must be treated so that it is safe. This has cost implications and some environmental implications, as water-treatment plants need to be built. Apart from the treatment of drinking water, where there is sanitation, sewage also needs to be treated before the water is recycled into the environment. Water polluted with human sewage is very dangerous to health. The details of treatment are given in the next section.

The public water supply

REVISED

In Britain, almost everybody gets water piped into their homes, and this water is fit to drink straight from the tap. The water originates mainly as rainfall, but it has to be cleaned and treated before it is safe to drink. The rainfall is collected in lakes or man-made reservoirs where it is stored. Alternative water supplies are rivers and underground water.

The water in reservoirs, lakes and rivers contains a number of things that need to be removed:
- large particles – lumps of soil, pebbles, etc.
- smaller suspended or floating particles – small soil particles, dead leaves, etc.
- microorganisms.

These impurities are removed in stages related to the size of the particles, from large to small. The process is shown in Figure 15.1.

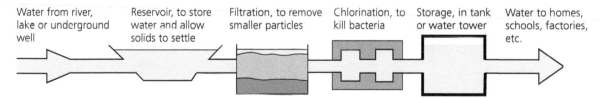

Water from river, lake or underground well | Reservoir, to store water and allow solids to settle | Filtration, to remove smaller particles | Chlorination, to kill bacteria | Storage, in tank or water tower | Water to homes, schools, factories, etc.

Figure 15.1 Stages in the treatment of drinking water.

1 **Sedimentation**. The larger insoluble particles are removed by simply allowing them to settle on the bottom of the reservoir or sedimentation tank.
2 **Filtration**. The smaller insoluble particles are removed by passing the water through a filter.
3 **Chlorination**. Chlorine is added. This is poisonous to many bacteria and so kills them. Tap water is not sterile and still does contain some bacteria, but the numbers are reduced and so are not a health hazard.

Now test yourself

TESTED

1 State three ways in which water can be conserved in the home.
2 What term describes the extraction of water for human use?
3 Apart from supplying drinking water, in what way would a supply of water benefit poor rural communities?
4 What is the purpose of sedimentation in the treatment of water?
5 What is added to water, during water treatment, to remove bacteria?

Answers online

Fluoridation, desalination and distillation

Fluoridation

In some areas of the UK, fluoride is added to drinking water. Fluoride is added in small amounts and is known to be helpful in preventing tooth decay. Tooth decay starts when the outer surface (enamel) of the tooth is attacked by acid produced by bacteria that live on the surfaces of the teeth in a layer known as plaque. These bacteria convert the sugars in our food into acid. The acid causes calcium and phosphates, of which enamel is largely composed, to dissolve. This process is called **demineralisation**. The saliva will eventually neutralise the acid and the minerals then re-enter the enamel – a process called **remineralisation**. Fluoride is beneficial in the following ways:

- In young children, who are developing their layer of enamel, fluoride alters the structure of the enamel so it is more resistant to acid attack.
- Low levels of fluoride encourage remineralisation.
- Fluoride reduces the ability of plaque bacteria to form acid.

The argument for adding fluoride to public water supplies is that it means that even those people who do not clean their teeth regularly or properly will get fluoride protection. The arguments against adding fluoride to public water include:

- A condition called **dental fluorosis** (white blotches on the enamel) can occur if teeth are exposed to high levels of fluoride. It does not damage the teeth nor cause any health problems.
- There is fluoride in around 90% of toothpastes, a much higher concentration than in fluoridated water, so some people argue that fluoridation is unnecessary.
- Some people believe that drinking water should be as natural as possible. They accept the addition of chlorine as a necessary safety measure, but believe that people should not be forced to take in any artificial additives like fluoride.

Desalination

We cannot drink sea water. The salt in it makes it taste unpleasant, but it also means that, if you did drink it, it would actually dehydrate you.

In some parts of the world, countries are looking at the possibility of removing the salt from sea water (**desalination**) so that it can be used for drinking. The usual process for desalination is called **reverse osmosis**. The sea water is filtered under pressure, which removes the salt. There are some problems with using desalination for supplying drinking water:

- The process uses a lot of energy because of the need to generate high pressures – much more than other processes that are used to produce drinking water.
- The process is expensive because of the energy needed.
- The desalination plants produce greenhouse gases, whereas normal water treatment plants produce very little.
- The very salty water left when the fresh water has been extracted is a pollutant and needs to be disposed of carefully.
- Some countries with a drought problem are very poor and could not afford desalination, while others do not have a coastline, so the water would have to be piped long distances.

Distillation

REVISED

Liquids that can mix together are called **miscible liquids**. **Distillation** is a technique that separates miscible liquids. Each chemical has its own boiling point, and for water this is 100°C. If you heat a mixture of liquids to 100°C, then any vapour coming off at that point is water. Other liquids will boil off at lower or higher temperatures.

In the laboratory, distillation is done with a piece of apparatus called a **condenser**, which is shown in Figure 15.2.

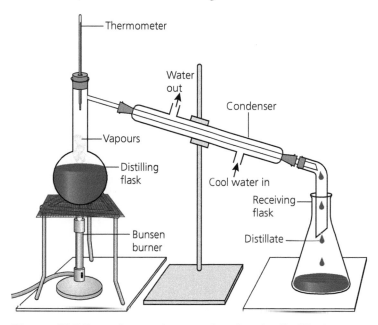

Figure 15.2 Experimental set-up for simple distillation.

Let us suppose we are trying to separate a mixture of alcohol (boiling point 78°C) and water (boiling point 100°C). When the liquid reaches 78°C, the alcohol will all turn into vapour (some will vaporise at lower temperatures, but at the boiling point it will vaporise completely). The vapour goes into the condenser, where the water running through the condenser condenses it back into a liquid. The alcohol then runs into the receiving flask, and the water is left in the distilling flask.

Solubility curves, hard and soft water

Solubility curves

REVISED

Water has been described as the 'universal solvent'. A wide variety of substances dissolve in it, to different extents. A chemical that dissolves in a solvent is called a **solute**. For each solute, there is a limit to how much can dissolve in a solvent. When no more can dissolve, the solution is said to be **saturated**. Solubility is measured in g/100 g solvent. Measurement involves gradually adding known amounts of solute until no more will dissolve. The solution is then filtered, dried and weighed to determine the excess of solute. By subtracting this excess from the total amount added, the weight of dissolved solute is obtained. Figure 15.3 shows the effect of temperature on the solubility of some sodium and potassium compounds. Such graphs are called **solubility curves**.

> **Exam tip**
>
> You do not need to learn the solubility curves shown in Figure 15.3. If an exam question involves solubility curves, these will be given. You do need to know the general principle that solubility increases with temperature.

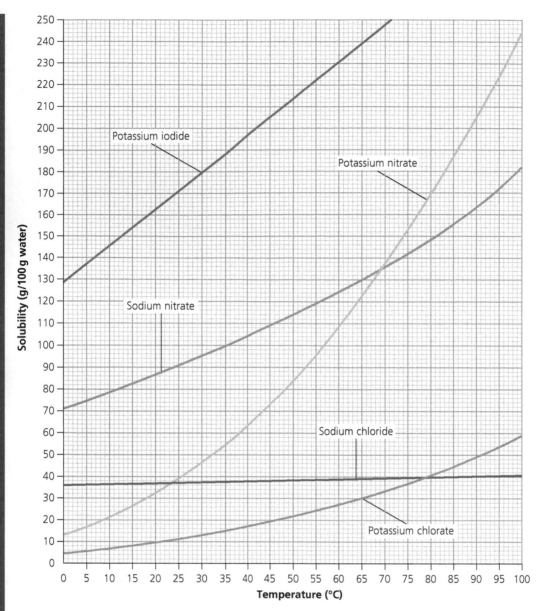

Figure 15.3 Solubility curves for some sodium and potassium compounds.

Hard and soft water

Tap water may be 'hard' or 'soft' depending on what area of the country you live in and where your water comes from. Hard water is water that contains dissolved calcium (Ca^{2+}) and magnesium (Mg^{2+}) ions. Hard water can be **temporary** or **permanent**, or a mixture of the two.

- **Temporary hard water** contains calcium hydrogencarbonate and/ or magnesium hydrogencarbonate. When this water is boiled, the hardness is removed, with the formation of solid calcium carbonate.

$$Ca(HCO_3)_2(aq) \rightarrow CaCO_3(s) + H_2O(l) + CO_2(g)$$

- **Permanent hard water** contains chlorides and/or sulfates of calcium and magnesium, and boiling does not soften this water.

The solid calcium carbonate formed when temporary hard water is heated can cause problems. Known as limescale, it can build up in hot water pipes and in appliances such as kettles and washing machines. However, there have been studies that suggest hard water may have health benefits, possibly decreasing the risk of heart disease.

Softening hard water

Hard water can be softened in three ways, which are summarised in Table 15.1.

Table 15.1 Methods of softening hard water.

Method	Used for	Advantages	Disadvantages
Boiling	Temporary hard water	Easy and cheap	Produces limescale. Only practical for small quantities. Only works on temporary hard water
Adding sodium carbonate	Temporary and permanent hard water	Cheap	Produces limescale. Only practical for small quantities
Ion exchange column	Temporary and permanent hard water	No limescale produced. Can treat large quantities of water	Expensive

Boiling to remove temporary hardness has been explained in the last section.

H Sodium carbonate (washing soda) is added to washing machine loads to soften the water. It prevents the calcium and magnesium ions bonding to the washing detergent, meaning that less detergent is used. The carbonate ions from sodium carbonate react with the calcium and magnesium ions in the water to produce **insoluble precipitates**.

$$Ca^{2+}(aq) + Na_2CO_3(aq) \rightarrow CaCO_3(s) + 2Na^+(aq)$$

An ion exchange column is a tube filled with a resin (Figure 15.4). Water is passed through the column and sodium ions on the resin are exchanged for calcium and magnesium ions, so removing them from the water and softening it. Eventually the resin has to be 'regenerated' by passing concentrated sodium chloride through it to replace the lost sodium ions.

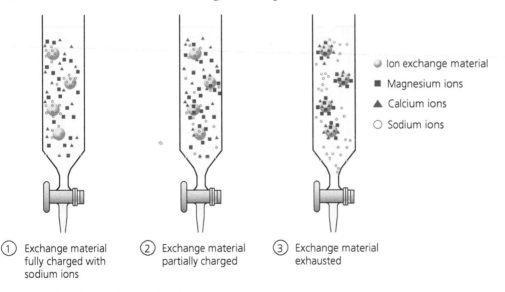

- ● Ion exchange material
- ■ Magnesium ions
- ▲ Calcium ions
- ○ Sodium ions

① Exchange material fully charged with sodium ions

② Exchange material partially charged

③ Exchange material exhausted

Figure 15.4 The water-softening process.

Now test yourself

6 What is the name of the dental condition which can be caused by an overdose of fluoride?
7 How can you tell if a solution is saturated with a given solute?
8 Which are the two metals whose salts cause water to be 'hard'?
9 If water contains calcium sulfate, will boiling soften it?
10 What is the chemical make-up of limescale?

Answers online

Summary

- Water in 'natural' water supplies contains dissolved gases, ions, microorganisms and pollutants.
- To achieve a sustainable water supply, certain measures need to be taken, including reducing our water consumption and reducing the environmental impacts of abstracting, distributing and treating water.
- The treatment of the public water supply uses sedimentation, filtration and chlorination.
- There are arguments for and against the fluoridation of the water supply in order to prevent tooth decay.
- Desalination of sea water can be used to supply drinking water but there are issues about the sustainability of this process on a large scale.
- Water and other miscible liquids can be separated by distillation.

- Solubility can be measured at different temperatures to produce a solubility curve.
- Hardness in water is caused by calcium and magnesium ions and it is possible to distinguish between hard and soft waters by their action with soap.
- Temporary hard water contains calcium and magnesium hydrogencarbonates; permanent hard water is caused by calcium and magnesium chlorides and/or sulfates.
- The processes used to soften water include boiling, adding sodium carbonate and ion exchange; there are advantages and disadvantages of different methods of water softening.
- **H** ● The way these methods work.
- Hard water has health benefits and harmful effects.

Exam practice

1 There have been reports that hard water may have health benefits and may be associated with a reduced risk of heart disease. It is thought that certain elements found in water, i.e. cadmium, sodium and lead, may raise blood pressure, which is a risk factor in heart disease. Another element, calcium, may protect the body from the effect of these metals.

A study in the USA compared the incidence of heart disease in four river valleys, two of which had soft water and two hard. The results were as follows.

River basin	Type of water	Death rate from heart disease (per 100,000)
Ohio river	Soft	39.7
Columbia river	Soft	33.3
Missouri river	Hard	28.3
Colorado river	Hard	23.3

a) What is the % increase in deaths from heart disease in the Ohio river basin compared to the Colorado river basin? [1]

b) What evidence is there from this study that heart disease deaths are lower in hard water areas? [2]

c) Suggest reasons why it would be unscientific to conclude from this study that hard water reduced the risk of heart disease. [2]

Another study, in Canada, compared the levels of the harmful metals cadmium and lead in hard and soft water. The results are shown in the table below.

Element	Estimated intake from food (µg)	Estimated intake from 2 dm³ of water per day (µg)	
		Soft water	Hard water
Cadmium	215	1.3	1.2
Lead	130	59.4	19.4

d) Suggest reasons why the cadmium intake in soft water areas is unlikely to be the cause of any increase in heart disease. [2]

→

e) From the information in the passage, suggest why hard water may act as a protective
measure against the effects of cadmium in the diet.   [2]

f) Suggest why the information presented here might suggest a possible disadvantage of using
ion exchange resins to soften hard water.   [2]

2 Water for human consumption undergoes three stages of treatment, **sedimentation**, **filtration**
and **chlorination**. In some areas a fourth stage, **fluoridation**, is carried out.

a) Explain the purpose of:
  i)   sedimentation   [1]
  ii) filtration   [1]
  iii) chlorination.   [1]

b) Some people do not agree with the addition of fluoride to our water supply.
  i) Give one advantage of having fluoride in the water.   [1]
  ii) State one reason why some people object to its addition.   [1]

c) In some parts of the world, sea water is treated to provide drinking water. Countries using
this process include Saudi Arabia, Israel and the USA.
  i) What name is given to the process of producing drinking water from sea water?   [1]
  ii) Suggest two reasons why it is not practical to carry out this process in many countries
of the world.   [2]

Answers and quick quiz 15 online

ONLINE

16 The ever-changing Earth

Structure and appearance

The structure of the Earth

The Earth has a structure composed of layers, as shown in Figure 16.1.

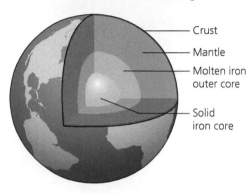

Crust
Mantle
Molten iron outer core
Solid iron core

Figure 16.1 The layers that make up the Earth.

- The **inner core** is the hottest part of the Earth, with a temperature of up to 5500°C. It is mostly iron, with some nickel.
- The **outer core** is a liquid layer, also made up of iron and nickel, with temperatures almost as high as the inner core.
- The **mantle** is the thickest layer of the Earth, consisting of semi-molten rock (more solid in the outer areas, more molten towards the centre).
- The **crust** is a thin layer of solid rock. Its actual thickness varies in different places, but it is up to 60 to 70 km thick. This compares to the mantle, which is about 2900 km thick.

Earth's changing appearance

Over the billions of years of its existence, the Earth has slowly but constantly changed its appearance. 200 million years ago, the land masses on Earth were all grouped together in one block, which scientists now call **Pangaea**. The continents can move across the surface of the planet.

Scientists now know that the surface of the Earth, or lithosphere, is made up of seven large **plates** and some smaller ones, about 70 km thick, which move a few centimetres per year. This movement is called **continental drift**.

The idea of continental drift was put forward by Alfred Wegener (1880–1930). He noticed that the continents of the Earth roughly fit together like a jigsaw. The coastlines of western Africa and eastern South America are a particularly good fit (see Figure 16.2). To explain this, some people had suggested that the continents may have moved, but there was no clear evidence for this apart from the 'jigsaw fit'. Alfred Wegener looked for evidence of continental drift, and found some:
- Rock formations on both sides of the Atlantic Ocean are exactly the same.
- Similar or identical animal and plant fossils are found in areas now separated by wide oceans.

- Certain fossils seem to be in the 'wrong place', e.g. fossil remains of semi-tropical species in northern Norway.

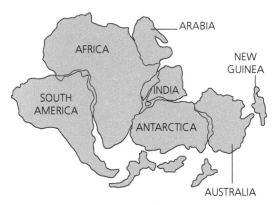

Figure 16.2 The continent 'jigsaw'. This group of continents is known as 'Gondwana'. The original land mass of Pangaea is thought to have split into two, Laurasia and Gondwana.

However, there was a weakness in Wegener's theory of continental drift. There was no known mechanism by which continents could plough their way through the Earth's crust without leaving any sort of 'trail'. Wegener's model was not accepted at the time, but new evidence gradually emerged and Wegener's theory has now been developed into the current theory of **continental drift**.

Plate tectonics and continental drift

REVISED ☐

New evidence that the continents could move across the surface of the Earth was found.
- Studies of the ocean floor found large mountain ranges and canyons. If the ocean floor was ancient it should have been smooth, because of all the sediment coming into it from rivers.
- In 1960, core samples taken from the floor of the Atlantic Ocean were analysed and dated. This showed that rock in the middle of the Atlantic was considerably younger than that from the eastern or western edges.
- No ocean floor was found to be older than about 175 million years, yet rocks on land had been found that were several billion years old.
- Rocks retain a record of the magnetic field of the Earth, which changes from time to time. Analysis of these magnetic records showed that Britain had spun round and moved north in the past, and that the patterns exactly matched those from North America.

It became clear that new ocean floor was forming all the time and spreading outwards, and in places near the borders of the continents it was sinking back into the crust. The Earth's crust was shown to be 'mobile'.

In the 1960s the idea of continental drift became generally accepted, and the theory was renamed **plate tectonics**.

The Earth's crust consists of solid plates, which fit together, with molten or semi-molten magma underneath. These plates can move independently of one another. Where two plates join, there can be three types of movement:
- The plates can collide at **convergent boundaries**. This 'crumples' the edges of the plates, forming mountain ranges. One plate may slide under the other. Magma is released and volcanoes can occur.

- The plates can move apart at **divergent boundaries** – molten rock (magma) below the surface is released. If this happens under pressure, this is a volcanic eruption.
- The plates can slide past one another at **conservative boundaries**, neither moving towards nor away from each other.

These movements are shown in Figure 16.3. Any movement of plates can lead to earthquakes. Plate movement causes a build-up of huge quantities of energy in the rock. When the energy is released, vibrations occur that travel through the rock, causing earthquakes of varying severity depending on the amount of energy built up.

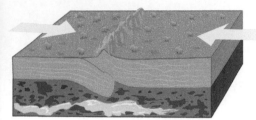
Convergent boundary

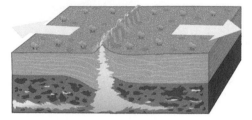
Divergent boundary

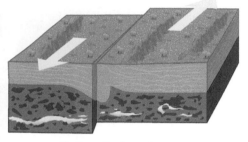
Conservative boundary

Figure 16.3 Types of plate movement.

Earth's atmosphere

Evolution of the atmosphere

REVISED

The original atmosphere of Earth was composed mainly of **hydrogen** and **helium**, but these low-density gases soon escaped Earth's gravity and drifted into space.

At this time, the Earth was young and was still cooling after its formation. There were large numbers of volcanoes on the surface, constantly erupting. The eruptions expelled a mixture of gases, including **water vapour**, **carbon dioxide** and **ammonia**. These gases built up in the atmosphere and the carbon dioxide dissolved in the early oceans, which were formed by the water vapour cooling and condensing, and from the ice contained in the millions of comets that were hitting Earth.

Eventually, simple plant life evolved in the oceans that could use the carbon dioxide and sunlight to make food by photosynthesis. **Oxygen** was released as a waste product, and so was added to the atmosphere. Oxygen allowed animal life to evolve, as the animals needed it for respiration.

The poisonous ammonia that was released from the volcanoes decomposed in sunlight to form **nitrogen** and hydrogen. The hydrogen escaped the atmosphere, but nitrogen remained, creating the atmosphere as it is today.

Some of the gases in the air can be extracted and used. These include **nitrogen, oxygen, neon** and **argon** (Table 16.1).

Table 16.1 Composition of the atmosphere today. The atmosphere also contains water vapour, but the amount varies.

Gas	Quantity in dry air, expressed in volumes
Nitrogen (N_2)	78.1%
Oxygen (O_2)	20.9%
Argon (Ar)	0.9%
Carbon dioxide (CO_2)	0.035%
Others:	0.065%
Neon (Ne)	
Helium (He)	
Krypton (Kr)	
Hydrogen (H_2)	
Ozone (O_3)	
Radon (Rn)	

Maintaining the atmosphere

REVISED

There is concern at the moment that the level of carbon dioxide in the atmosphere is going up. This is a concern because it is new – the proportions of the gases have remained constant for millions of years. This is despite living organisms using and producing both oxygen and carbon dioxide. In the past, the processes using oxygen have been balanced by those producing it, and the same has applied to carbon dioxide. Oxygen production and use is still balanced, but carbon dioxide production is now exceeding use.

Oxygen

Nearly all living things use oxygen to obtain energy from respiration. Oxygen is produced by plants during photosynthesis and, because they produce more than they need for their own respiration, they add it back to the atmosphere. In terms of oxygen, respiration and photosynthesis balance each other (Figure 16.4).

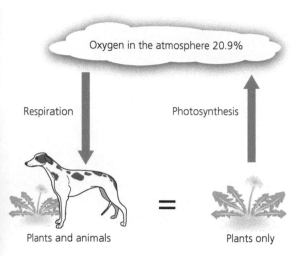

Figure 16.4 The balance of oxygen use and production.

Carbon dioxide

The carbon cycle is summarised in Figure 16.5. Under certain conditions the dead bodies of plants and animals become fossilised rather than decaying, and the carbon in their bodies remains fixed inside them rather than being released back into the atmosphere as carbon dioxide. In this way, fossil fuels (oil, coal and natural gas) have stored carbon for millions of years. In the last 150 years, the discovery of oil and natural gas and the huge growth of industry has seen an enormous increase in the burning of fuels, and carbon that has taken millions of years to build up in the Earth has been released rapidly into the atmosphere in the form of carbon dioxide. This **combustion** (burning) of fossil fuels has disturbed the balance that existed before and the level of carbon dioxide in the atmosphere has increased instead of remaining constant. This change is thought to have caused changes in the Earth's climate.

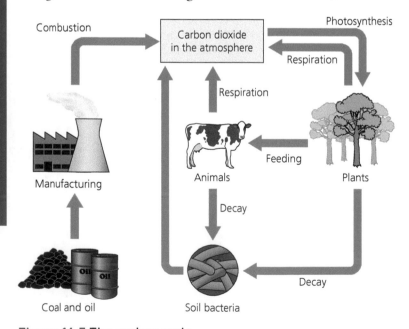

Figure 16.5 The carbon cycle.

Acid rain

The burning of fossil fuels is a major cause of **acid rain**. Sulfur dioxide and nitrous oxides (produced by burning fossil fuels) are acidic gases that react with water to form acids.

oxides of nitrogen + water → nitric acid

sulfur dioxide + water → sulfuric acid

These gases dissolve in the water vapour in the atmosphere and condense into clouds, which then produce acid rain. The acid rain can kill wildlife, with fish and conifer trees being particularly affected. The acid rain also causes damage to limestone buildings because acids react with limestone, causing it to dissolve in water.

To reduce acid rain, we need to reduce fossil fuel use and fit devices known as 'scrubbers' to factory chimneys. These remove sulfur and nitrous oxides from the waste gases. More details are given in the next section.

Now test yourself

6 Which two gases made up most of Earth's original atmosphere?
7 How did the activity of volcanoes play a part in the formation of the first oceans?
8 What was the cause of oxygen first appearing in the atmosphere?
9 How do animals get the carbon necessary to build their bodies?
10 Which gases are the main cause of acid rain?

Answers online

Earth's atmosphere 2

Managing change in the atmosphere

Global warming could have grave consequences for the Earth and its inhabitants.
● It may lead to more extreme (and dangerous) weather.
● Melting of the ice caps would cause rising sea levels and flooding of large areas of land.
● Climate change could cause the destruction of habitats.

In order to reduce the problems of increasing carbon dioxide levels in the atmosphere, certain measures can be taken.

Reducing the use of fossil fuels

Measures can be taken by governments, industry and individuals to reduce the burning of fossil fuels. These include:
● Use of nuclear and renewable power rather than power provided by burning coal, oil and gas.
● Recycling or reusing materials as much as possible, so that less fossil fuels are used to make replacements.
● Developing and using more fuel-efficient vehicles.
● Reducing energy consumption in the home by various means (e.g. insulating homes efficiently, lowering central heating temperatures by a degree or two, using low-energy light bulbs, not leaving televisions and game consoles on standby, using and regularly servicing fuel-efficient boilers, etc.).
● Using means of mass transportation (i.e. trains and buses) rather than personal vehicles, or sharing cars where possible.

Boosting the natural removal of carbon dioxide

Plants remove carbon dioxide from the atmosphere. The huge areas of forest being destroyed for commercial reasons and to clear land for farms or habitation means that the Earth is losing a lot of its capacity to absorb carbon dioxide from the atmosphere. Measures that can help here include:

- drastically reducing deforestation
- re-forestation, i.e. planting trees in large numbers.

Cleaning fuel emissions

It is possible to remove some of the harmful gases produced by the burning of fossil fuels before they enter the atmosphere. At the moment it is only practical to do this on a large scale, such as in power plants.

Carbon capture can reduce the carbon dioxide emissions from power stations by around 90%. It is a three-step process:

- capturing the CO_2 from power plants and other industrial sources
- transporting it, usually via pipelines, to storage points
- storing it safely in geological sites such as depleted oil and gas fields.

The most commonly used form of carbon capture is post-combustion capture, which involves capturing the carbon dioxide from the gases given off by burning. A chemical solvent is used to separate carbon dioxide from the waste gases.

There are also techniques being developed for removing the sulfur dioxide from the waste gases produced by power stations. Such processes are referred to as **sulfur scrubbing**, and can reduce the levels of sulfur dioxide by more than 95%.

Testing for atmospheric gases

Table 16.2 details the tests that you need to know for different atmospheric gases.

Table 16.2 Tests for gases.

Gas	Tested by	Positive result
Oxygen	Glowing splint	Splint relights
Carbon dioxide	Bubble through lime water	Lime water goes cloudy
Hydrogen	Lighted splint	Gas burns with a 'squeaky pop'

Now test yourself

TESTED

11 How could global warming lead to the destruction of habitats?
12 How can recycling help to reduce global warming?
13 How does deforestation contribute to global warming?
14 What is a sulfur scrubber?
15 What is the test for oxygen?

Answers online

Summary

- The Earth has a solid iron core, molten iron outer core, a mantle and a crust.
- The theory of plate tectonics was developed from Alfred Wegener's earlier theory of continental drift.
- At tectonic plate boundaries, plates may slide past one another, move towards one another or move apart.
- The formation of the Earth's first true atmosphere occurred when gases, including carbon dioxide and water vapour, were expelled from volcanoes.
- The present atmosphere consists of nitrogen, oxygen, argon, carbon dioxide and traces of other gases.
- The air is a source of nitrogen, oxygen, neon and argon.
- The composition of the atmosphere has changed over geological time.
- The processes of respiration, combustion and photosynthesis maintain the levels of oxygen and carbon dioxide in the atmosphere.
- The emission of carbon dioxide and sulfur dioxide into the atmosphere through the combustion of fossil fuels causes global warming and acid rain.
- Certain measures can be used to try to address the problems of global warming and acid rain.
- Oxygen relights a glowing splint.
- Carbon dioxide turns lime water cloudy.
- Hydrogen burns with a squeaky pop.

Exam practice

1 The diagram shows a tectonic plate boundary.
   a) What type of plate boundary is shown? [1]
   b) Name layer A. [1]
   c) Name the molten material B. [1]
   d) Studies have shown that the rock in the middle of the Earth's oceans where plate boundaries are found is up to 140 million years 'younger' than the rocks at the edges of the oceans. Explain how this supports the tectonic theory. [4]

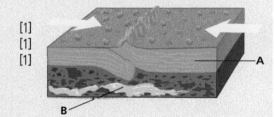

Figure 16.6

2 Sulfur dioxide is a pollutant in waste gases from power stations. It is a cause of acid rain. The sulfur dioxide can be removed by a process known as flue gas desulfurisation, in which the sulfur dioxide is neutralised by calcium oxide to form calcium sulfate. The equation for the process is shown below.

$$CaO(s) + SO_2(g) \rightarrow CaSO_3(s)$$

   a) How does the release of sulfur dioxide result in the formation of acid rain? [3]
   b) Which other gas is a main contributor to the formation of acid rain? [1]
   c) What is the minimum amount of calcium oxide that would be needed to neutralise 100 g of sulfur dioxide? [3]
      (Relative atomic masses: Ca = 40; S = 32; O = 16.)
   d) State **two** ways in which acid rain is harmful to the environment. [2]

Answers and quick quiz 16 online

ONLINE

17 Rate of chemical change

Measuring the rate of a reaction

The rate of a reaction means how much product (usually by mass or volume) is produced in a set time (usually per second). A graph of amount of product against time allows us to determine the rate of the reaction – the gradient (slope) of the graph tells us how much product is being produced per unit time. There are three simple ways of measuring rates of reaction in a school laboratory.

Measuring the volume of gas produced

This can be done by displacing water from a graduated tube or by collecting the gas directly using a gas syringe. The volume of gas produced will increase with time, up to a maximum volume.

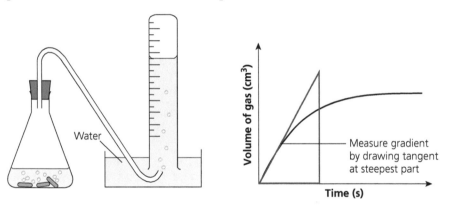

Figure 17.1 A method of capturing gas produced, and finding the rate of a reaction from the gradient of a volume–time graph.

Measuring the amount of light passing through

Some reactions produce a solid precipitate. If light is shone through the reaction, the light intensity passing through the reaction will decrease as more and more precipitate forms. This can be measured using a light sensor.

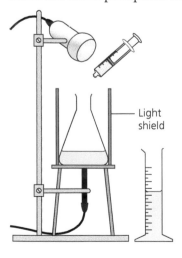

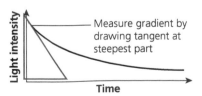

Figure 17.2 Measuring the intensity of light passing through a reaction vessel, and finding the rate of the reaction from the gradient of a intensity–time graph.

Measuring the change in mass

An electronic balance can be used to measure the overall mass of a reaction. If gas is produced as one of the reaction products, then the overall mass of the reaction will decrease with time as the gas escapes.

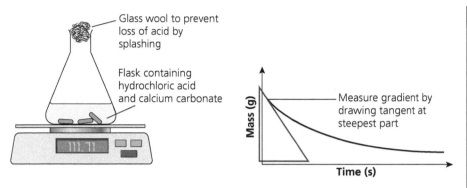

Figure 17.3 Measuring the mass of a reaction, and finding the rate of the reaction from the gradient of a mass–time graph.

Exam tip

Rate of reaction questions frequently involve plotting/sketching and analysing graphs. Make sure that you read the graph axes carefully – check the units and the scale. If you have to give a value from the graph in the exam, use a ruler and a sharp pencil to read off values from the axes.

Now test yourself

1 A student studied the rate of a reaction between hydrochloric acid and marble chips. He added excess acid to different amounts of marble chips and recorded the volume of gas produced every minute. The results are shown in the graph.

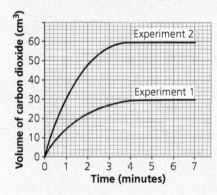

Figure 17.4

a) Use the graph to answer the following questions.
    i) Give the volume of gas produced after 1 minute in Experiment 2.
    ii) Give the time taken for the reaction to end in Experiment 1.
b) 0.2 g of marble chips were used in Experiment 1. Give the mass of marble chips used in Experiment 2.
c) The reaction in Experiment 2 was faster than that in Experiment 1. Explain how the graph shows this.

2 The graph shows the volume of carbon dioxide produced when excess limestone is added to 100 cm³ of hydrochloric acid at room temperature.

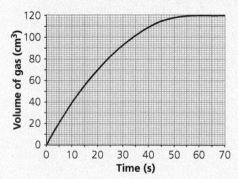

Figure 17.5

a) Use the graph to find:
    i)   the volume of carbon dioxide produced after 20 seconds
    ii) the time taken for the reaction to stop.
b) On the graph, draw the curve you would expect if the reaction were repeated using exactly the same volume and concentration of acid at a higher temperature, with the limestone still in excess.

Answers online

Explaining the rate of a reaction

A chemical reaction occurs when the reacting particles collide with one another. Not all collisions result in a chemical reaction: the collision needs enough energy for bonds in the reactant molecules to break. In Figure 17.6, particles of two different reacting gases are shown in yellow and red. All the particles are moving at high speed in random directions. When one of the red particles collides with one of the yellow particles with enough energy, then a reaction will occur. The number of successful collisions is a small percentage of the total number of collisions taking place in a given time. The number of successful collisions per second is called the collision frequency. The higher the collision frequency, the higher the rate of reaction.

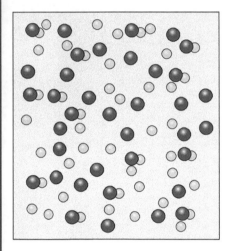

Figure 17.6 When particles of reacting gases collide with sufficient energy a chemical reaction will occur.

Factors affecting the rate of chemical change

REVISED

The temperature of the reactants

A higher temperature means a higher rate of reaction. This is because:
● increasing the temperature of the reactants increases the mean velocity of the particles
● faster particles mean that more of the collisions will have enough energy for a reaction to take place – the collision frequency increases
● a higher collision frequency means a higher rate of reaction.

The concentration of the reactants

A higher concentration means a higher rate of reaction. This is because:
● increasing the concentration of the reactants increases the total number of reacting particles in the same volume
● more particles per unit volume means there are likely to be more collisions with enough energy and so the collision frequency increases
● a higher collision frequency means a higher rate of reaction.

The surface area of the reactants

A higher surface area means a higher rate of reaction (Figure 17.7). This is because:

- increasing the surface area allows more of the reactants to collide with each other with enough energy
- more collisions cause a higher collision frequency and so a higher rate of reaction.

Using a catalyst

A catalyst can also increase the rate of reaction.

- Catalysts are substances that increase the rate of a chemical reaction but remain chemically unchanged at the end of the reaction.
- Many catalysts work only for one particular reaction.
- Some industrial reactions are only possible on large scales by using a catalyst.
- A catalyst works because it provides a 'surface' on which the reacting molecules can collide with each other.
- **H** This reduces the amount of energy needed for a collision to be successful.
- More particles will have the minimum amount of energy required, increasing the collision frequency and the rate of reaction.

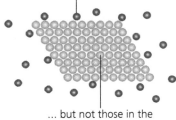

Red particles can hit the outer layer of green particles

... but not those in the centre of the lump

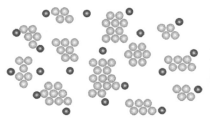

With the same number of green particles now split into lots of smaller bits, there are hardly any green particles that the red particles can't get at

Figure 17.7 Powders react at a faster rate than lumps.

The importance of catalysts

REVISED

Catalysts are used in the production of bulk chemicals, fine chemicals, petrochemicals and in food processing: 90% of all commercial chemical products involve catalysts at some stage in their manufacture. The use of catalysts in chemical processes has huge importance, not only in economic terms. Catalysts reduce the amount of energy required to produce chemical products. This, in turn, preserves world fuel reserves and reduces the environmental impact of burning fossil fuels, e.g. global warming.

> **Exam tip**
>
> When you have to plot a graph in an exam, you usually have to devise a suitable scale for each axis, plot the points correctly and draw a suitable line of best-fit. Make sure that your axes have linear scales. Check that you have plotted all the points correctly. Remember that best-fit lines can be curves as well as straight lines.

Now test yourself

TESTED

3 a) Fillers can be used to fill dents in car bodies. These work by mixing a paste with a small quantity of hardener. The paste does not harden until the hardener is added. The hardener acts as a catalyst for the reaction. State what is meant by a catalyst.

  b) A student wanted to find out if the amount of catalyst made a difference to the rate at which the paste hardened. She mixed a fixed amount of paste with different volumes of catalyst and recorded the time it took for the paste to harden at room temperature, 20°C. The results obtained are given in the table.

→

Volume of hardener added to paste (cm³)	Time taken for past to harden (min)
0.5	15.0
1.0	7.5
1.5	5.0
2.0	3.5
2.5	2.0
3.0	2.0

i) On a piece of graph paper, draw a graph of the results.
ii) Using your graph, describe what the results tell you.
iii) State what you would expect to happen to the reaction rate if the experiments were repeated at 50°C.

Answers online

Summary

- Plan and carry out experiments to study the effect of any relevant factor on the rate of a chemical reaction, using appropriate technology e.g. a light sensor and datalogger to follow the precipitation of sulfur during the reaction between sodium thiosulfate and hydrochloric acid.
- Analyse data collected in order to draw conclusions, and critically evaluate the method of data collection, the quality of the data and to what extent the data support the conclusion.
- Explore the particle theory explanation of rate changes, arising from changing concentration

(pressure), temperature and particle size, using a range of sources including textbooks and computer simulations.
- A catalyst increases the rate of a chemical change while remaining chemically unchanged itself and reduces the energy required for a collision to be successful.
- There is great economic and environmental importance in developing new and better catalysts, in terms of increasing yields, preserving raw materials, reducing energy costs, etc.

Exam practice

1 Trystan carried out an investigation into the reaction between dilute hydrochloric acid (HCl) and magnesium ribbon. He reacted the magnesium with five different concentrations of acid and measured the volume of hydrogen gas produced after 30 s using the apparatus below.

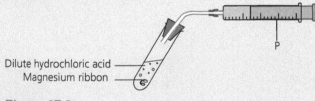

Dilute hydrochloric acid
Magnesium ribbon

Figure 17.8

a) Name apparatus P. [1]

b) Trystan's results are shown below.

Concentration of HCl (mol/dm³)	Volume of H₂ gas produced (cm³)
0.2	8
0.5	17
1.0	26
1.5	30
2.0	30

i) State what can be concluded about the effect of concentration of acid on the rate of the reaction. Explain this effect using your understanding of particle theory. [3]

ii) Trystan initially measured the volume of gas collected in 60 s. Explain why he amended his plan after making these measurements. [2]

iii) State two factors other than concentration which could affect the rate of the reaction between hydrochloric acid and magnesium. [2]

c) Limestone is made of calcium carbonate. It reacts slowly with acid rain and is gradually eaten away. Design an experiment based on this reaction to identify which of three samples of rainwater is the most acidic. [3]

(From WJEC GCSE Chemistry Unit 1 Foundation Tier Sample Assessment Materials Q10)

2 Hydrogen peroxide decomposes into water and oxygen. This reaction is catalysed by metal oxides.

hydrogen peroxide → water + oxygen

a) State the purpose of using a catalyst. [1]

b) In an investigation, 2 g of three different metal oxides were added to 25 cm³ of hydrogen peroxide containing two drops of washing-up liquid. The washing-up liquid formed a froth as oxygen was given off.

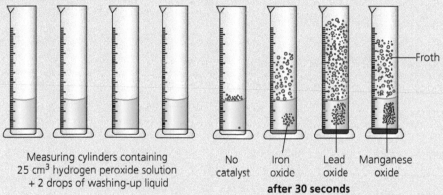

Measuring cylinders containing 25 cm³ hydrogen peroxide solution + 2 drops of washing-up liquid

No catalyst   Iron oxide   Lead oxide   Manganese oxide   ──Froth

after 30 seconds

Figure 17.9

i) State the best catalyst. Give the reason for your choice. [2]

ii) The following factors were kept the same in the investigation: mass of metal oxide; volume and concentration of hydrogen peroxide; amount and type of washing-up liquid. Give one other factor that must be kept the same to make the investigation a fair test. [1]

iii) After 30 minutes the reaction had stopped in all of the cylinders containing catalysts. Give the reason for this. [1]

iv) When the reaction had stopped all the remaining catalyst from one of the measuring cylinders was recovered and dried. Which of the following answers, A, B or C, shows the mass which remained? [1]

   A   More than 2 g

   B   2 g

   C   Less than 2 g

(From WJEC GCSE Chemistry 2 Foundation Jan 17 Q4)

3 Sodium thiosulfate solution reacts with dilute hydrochloric acid forming a yellow precipitate. The yellow precipitate formed during the reaction causes a cross marked on a piece of white paper to disappear. The time taken for this to happen is measured. 10 cm³ of dilute hydrochloric acid were added to 50 cm³ sodium thiosulfate solutions of five different concentrations. The results are shown below.

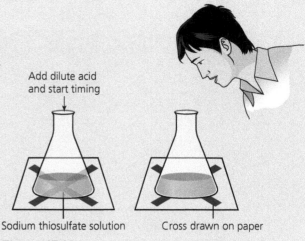

Add dilute acid and start timing

Sodium thiosulfate solution   Cross drawn on paper

Figure 17.10

Concentration of sodium thiosulfate solution (g/dm³)	Time for cross to appear (s)		
	1	2	Mean
8	53	55	54
16	31	31	31
24	19	21	20
32	14	22	18
40	12	12	12

a) i) State which concentration gave the least accurate mean time. [1]
   ii) Explain your choice of concentration in part (i) and suggest a practical reason which might have caused an unexpected result. [2]

b) A graph was plotted for the mean time for the cross to disappear against the concentration of sodium thiosulfate solution.
   i) Use the graph to find the mean time for the cross to disappear when sodium thiosulfate solutions of 10 g/dm³ and 20 g/dm³ are used. Complete the table below. [1]

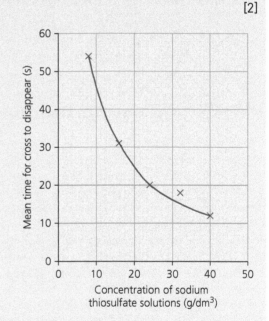

Figure 17.11

Concentration of sodium thiosulfate solution (g/dm³)	Mean time (s)
10	
20	
40	12

   ii) Use the information from part (i) to describe how the mean time changes with increasing concentration of sodium thiosulfate. [2]

c) The equation shows the reaction occurring in the flask.

$Na_2S_2O_3(aq) + 2HCl(aq) \rightarrow 2NaCl(aq) + SO_2(g) + S(s) + H_2O(l)$

State which product is the yellow precipitate. [1]

d) This method needs a person to view the formation of the precipitate. Give the name of a piece of equipment that could be used to monitor the reaction instead of a person. Describe what it measures. [2]

(From WJEC GCSE Chemistry 2 Foundation Jan 17 Q5)

4 The following graphs show the volume of hydrogen produced over time during the reaction between magnesium and hydrochloric acid of two different concentrations. All other factors were kept constant.

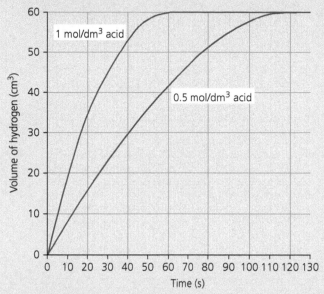

Figure 17.12

a) State what conclusion can be drawn from the graph and use your understanding of particle theory to explain that conclusion. [4]

b) Another method of studying this reaction is to use a balance to record the change in mass over time. The data can be recorded directly on a computer.

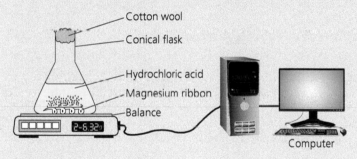

Figure 17.13

i) State why a two decimal place balance is required for this method to work. [1]

ii) Use the relative atomic mass values below to explain why recording the change in mass is better suited to an experiment that releases carbon dioxide, CO_2, than one that releases hydrogen, H_2. [3]

$A_r(H) = 1$; $A_r(C) = 12$; $A_r(O) = 16$

(From WJEC GCSE Chemistry 2 Higher Jun 16 Q5)

Answers and quick quiz 17 online

ONLINE

18 Bonding, structure and properties

Metals and ionic compounds

The properties of metals

The metals have similar **physical** and **chemical properties**, which distinguish them from non-metals.

The general properties of metals are as follows:
- **Strong**: A 'strong' material is one that cannot be easily broken. The opposite of strong is brittle. Metals are generally strong, although some are much stronger than others.
- **Malleable and ductile**: Malleable means that the material can be bent, hammered or squashed into different permanent shapes. Ductile means that the material can be stretched out into wires.
- **High melting and boiling points**: An exception to this is mercury, which is a liquid at room temperature.
- **Good conductors of electricity**: All metals conduct electricity to some extent, although their conductivity varies considerably.
- **Good conductors of heat**: The metals have a high **thermal conductivity**. Those that are better conductors of electricity are also better conductors of heat.
- **Shiny when polished**: This is sometimes described as being **lustrous**. The more reactive metals tend to build up a layer of oxide on their surface (e.g. rust on iron) which dulls them.
- **High density**: The way that metal atoms bond together tends to give them a high density.
- **React with oxygen to form basic oxides**: the reactivity of metals varies quite a lot. Platinum, gold and silver are the least reactive metals and do not naturally form oxides.

> **Exam tip**
>
> Although all metals are good conductors of electricity, it is not true that all non-metals are not. Carbon, a non-metal, is a very good conductor of electricity (when in the form of graphite).

The conductivity of metals

Copper is a particularly good conductor of electricity and heat, and is used as an example here.
- Like all solid metals, copper has a structure consisting of a **lattice** (regular three-dimensional pattern) of positive ions through which a 'sea' of **free electrons** move.
- The positive ions and the 'sea' of electrons interact to form metallic bonds. The free electrons are the outermost electrons on the copper atoms, which are stripped from the atoms when they 'freeze' together to make the solid copper; the rest of the atom forms the positive ion cores.
- The lattice / free electron model explains the high electrical and thermal conductivity of metals including copper. The sea of electrons can easily move throughout the structure of the metal. Electrons carry a negative charge, so electric current is the movement of the free electrons through the structure – from negative to positive.

- The positive ions are close together and bonded by metallic bonds.
- The structure can easily pass the vibration of hot particles from one particle to the next particle; the free electrons can move faster as they are heated, and transfer the heat throughout the structure – this explains why metals are good thermal conductors.
- The arrangement of the positive ions in copper and the number of free electrons makes copper a particularly good conductor of electricity and heat (see Figure 18.1).

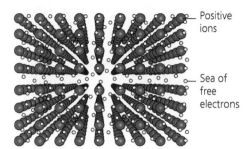

Positive ions

Sea of free electrons

Figure 18.1 The metallic structure of copper.

Ionic compounds

REVISED ☐

Ionic compounds are ones in which the particles are joined by **ionic bonds**. Ionic bonds are formed between oppositely charged particles (**ions**). Atoms are most stable when they have a full outer shell of electrons. Atoms with relatively full outer shells (which will be non-metals) tend to gain electrons and so acquire a negative charge. Atoms with very few electrons in their outer shell (these will be metals) will tend to lose those electrons and become positively charged. Positively and negatively charged particles attract one another. If a metallic and a non-metallic ionic element are close together, the metal will donate one or more electrons to the non-metal, and the two charged ions will then bond together. Figure 18.2 illustrates this process for the formation of sodium chloride.

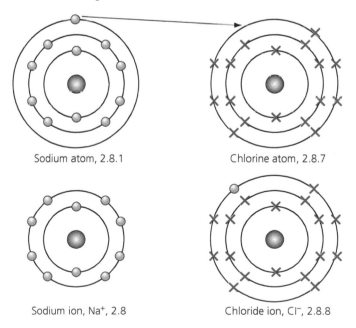

Sodium atom, 2.8.1   Chlorine atom, 2.8.7

Sodium ion, Na$^+$, 2.8   Chloride ion, Cl$^-$, 2.8.8

Figure 18.2 Sodium and chlorine atoms and ions.

When sodium reacts with chlorine to make sodium chloride, the sodium atom will need to lose its outermost electron (to allow its new outer shell to be full), and chlorine will need to gain an electron (to fill its outer shell). The sodium atoms will become sodium ions, Na^+ (because they have lost an electron), and the chlorine atoms will become chloride ions, Cl^- (because they have gained an electron). There will be a strong electrostatic attraction between the oppositely charged ions, which is called an **ionic bond**.

- Ionic bonds are formed by the transfer of electrons from one atom to another.
- The salts which are formed when metals react with non-metals have names that reflect the metal ion and the non-metal ion. The metal ions have the same name as their atom, so sodium atoms become sodium ions. Non-metal ions have slightly different names. Oxygen forms oxides, fluorine forms fluorides, chlorine forms chlorides, bromine forms bromides and iodine forms iodides.
- The formulae of simple ionic compounds can be written using the formulae of the ions given in Tables 18.1 and 18.2. Whenever ionic compounds are formed from metals and non-metals, the resulting compounds are electrically neutral – the number of positive charges balances the number of negative charges.

The ions formed allow us to work out the formula of an ionic compound. For example, calcium chloride. The calcium ion is Ca^{2+}. This means that two chloride ions are needed to balance the one calcium ion. The formula for calcium chloride is $(Ca^{2+})(2Cl^-)$, which we write as $CaCl_2$.

Table 18.1 Formulae of some positive ions.

Charge: +1		Charge: +2		Charge: +3	
Sodium	Na^+	Magnesium	Mg^{2+}	Aluminium	Al^{3+}
Potassium	K^+	Calcium	Ca^{2+}	Iron(III)	Fe^{3+}
Lithium	Li^+	Barium	Ba^{2+}	Chromium(III)	Cr^{3+}
Ammonium	NH_4^+	Copper(II)	Cu^{2+}		
Silver	Ag^+	Lead(II)	Pb^{2+}		
		Iron(II)	Fe^{2+}		

Table 18.2 Formulae of some negative ions.

Charge: −1		Charge: −2		Charge: −3	
Chloride	Cl^-	Oxide	O^{2-}	Phosphate	PO_4^{3-}
Bromide	Br^-	Sulfate	SO_4^{2-}		
Iodide	I^-	Carbonate	CO_3^{2-}		
Hydroxide	OH^-				
Nitrate	NO_3^-				

Now test yourself

TESTED ☐

1 Metals are good conductors of electricity. Why is it useful (in that context) that they are also ductile?
2 Which types of metals will be relatively less lustrous?
3 Why are metals good conductors of electricity?
4 Why does sodium form ions easily?
5 Look at Tables 18.1 and 18.2. What will be the formula of barium hydroxide?

Answers online

Ionic and covalent compounds

Properties of ionic compounds

Sodium chloride is a typical example of an ionic compound. Its properties are as follows.

- It has a three-dimensional lattice structure (a regular, repeating structure) held together by strong electrostatic forces (see Figure 18.3). This is referred to as a **giant ionic structure**.
- The ions in sodium chloride crystals are arranged in a **cubic lattice**, each ion being surrounded by six nearest neighbours of opposite charge. This is shown in Figure 18.3. The simplest way of drawing the arrangement of the ions in the lattice is shown in Figure 18.4 – this is called the **unit cell** of sodium chloride.
- It has a **high melting point**. Each sodium ion attracts all the chloride ions around it, and *vice versa*, so a lot of energy is needed to overcome the attractive force. This energy can be provided by heating, but high temperatures are needed to supply enough energy to break the ionic bonds.
- It **conducts electricity when molten or dissolved in water**. Solid sodium chloride does not conduct electricity because the sodium and chloride ions are held in fixed positions within their lattice and are not free to move. If there are no moving charged particles, there will be no electrical conduction. When molten or dissolved in water, the lattice breaks down, and the ions are free to move – the moving ions create an electrical current.
- **Brittleness** is a typical property of ionic substances. If a stress is applied to a crystal, which shifts the ion layers slightly, the ion layers will tend to jump over each other. Ions of the same charge are then brought side by side and so repel each other and the crystal fractures.

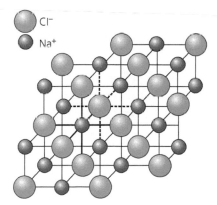

Figure 18.3 Sodium chloride lattice of ions.

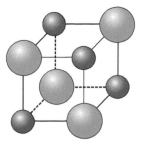

Figure 18.4 The unit cell of sodium chloride.

Covize bonds and covalent compounds

It is very difficult for some atoms to fill their outer shell by gaining or losing electrons. Once one electron is lost, for example, the resulting positive charge will tend to attract electrons, and so make it difficult to lose another one. Atoms that need to lose or gain three or four electrons to have a full outer shell rarely or never form ions. They form compounds by forming **covalent bonds**, in which **electrons are shared**. Water is an example of a covalent compound.

When two or more atoms join together by covalent bonds, the resulting structure is called a **molecule**.

When water molecules form as hydrogen reacts with oxygen, two atoms of hydrogen combine with one atom of oxygen. Hydrogen has one electron in its outer shell, and needs one more to get two to fill its outer shell. Oxygen has six electrons in its outer shell and needs to get two more to fill its outer shell.

Figure 18.5 shows the electron dot and cross diagram for hydrogen and oxygen forming water. The oxygen atom and the two hydrogen atoms share electrons – each atom then has a full outer electron shell.

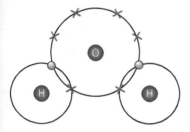

Figure 18.5 A water molecule showing how the covalent bonds are formed.

The covalent bond is a pair of electrons shared between two atoms. Covalent bonds are very strong. It takes a lot of energy to break them. The molecules themselves, however, attract each other only very weakly as each molecule is overall neutral. This means that solid covalent substances have **low melting points**, because very little energy is required to separate the molecules and turn the solid to a liquid. The weakness of these intermolecular forces explains why so many covalent compounds like water are liquids or gases at room temperature.

Covalent bonds are usually represented by a line between the atoms, illustrating where the bond is. This type of diagram is called a **structural formula**, and although the molecules themselves are three-dimensional, structural formulae are normally drawn in two dimensions. The structural formula of water is shown in Figure 18.6.

Figure 18.6 Structural formula of water.

Properties of covalent compounds

- They have **low melting and boiling points**. At room temperature, most covalent compounds are gases or liquids.
- They **do not conduct electricity** because they have no free electrons and do not form ions, so they cannot carry a current.

Giant covalent compounds and their properties

Giant covalent compounds

REVISED

Some covalent substances exist as giant structures that have high melting points because the atoms are held together by very strong covalent bonds. Although covalent substances normally have low melting and boiling points because of weak intermolecular forces, giant covalent substances are, in effect, one large molecule.

Graphite and diamond are examples of giant covalent structures. Diamond and graphite are both different physical forms of carbon. Both diamond and graphite contain covalent bonds between carbon atoms (Figures 18.7 and 18.8).

The carbon atoms in diamond are each connected to four other carbon atoms by a strong covalent bond. The structure is a three-dimensional lattice based on a tetrahedral unit cell, each carbon atom at the corner of a tetrahedron (Figure 18.9).

Graphite has very different properties to diamond, although both are made solely of carbon atoms joined by covalent bonds (see Table 18.3). This is because the atoms are arranged in a different way. Graphite is made up of layers of carbon atoms, arranged in hexagonal rings (Figure 18.10). Each carbon atom forms a strong covalent bond with three others in the same layer. However, the bonds between the layers are quite weak and allow the hexagonal-ring layers to slide over one another.

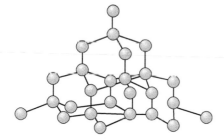

Figure 18.7 The structure of diamond.

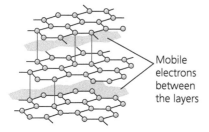

Mobile electrons between the layers

Figure 18.8 The structure of graphite.

Figure 18.9 Tetrahedral diamond unit cell.

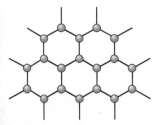

Figure 18.10 Hexagonal layer structure of graphite.

Table 18.3 Physical properties of diamond and graphite.

Physical properties of diamond	Physical properties of graphite
Transparent and crystalline – used as a gemstone in jewellery	Grey/black shiny solid
Extremely hard – used for glass cutting, small industrial diamonds are used in drill bits for oil exploration etc.	Very soft – used as a lubricant and used in pencils (the softer the pencil, the more graphite there is in the 'lead')
Electrical insulator	Is a non-metal that conducts electricity. Graphite is used for electrodes in some manufacturing processes
Very high melting point, over 3500°C	Very high melting point, over 3600°C

Diamond and graphite – linking structure and properties

REVISED

The carbon atoms in the layers of graphite are held together by three strong covalent bonds, forming the hexagonal-ring layer. Carbon is a Group 4 element and so it has four outer electrons; it needs to share four other electrons to get a complete outer electron shell. It gets three of these electrons from the carbon atoms in the hexagonal ring. The fourth electron from each atom, which is not used in bonding within the layers, joins a delocalised system of electrons between the layers of carbon atoms. Graphite conducts electricity along the layers because it has electrons that are free to move, forming an electric current. Graphite does not conduct electricity across the layers. The hexagonal layers in graphite can slide over one another (because the bonds between the layers are very weak), which gives it a slippery feel and lubricating properties. The carbon atoms are held together by strong covalent bonds, so it has a high melting point.

In diamond, all four of the outer electrons are involved in covalent bonding with four other carbon atoms. The result is a giant rigid covalent structure. This gives diamond its incredible hardness and high melting point, as a lot of energy is required to break down the lattice. There are no free electrons to conduct electricity.

Exam tip

Exam questions dealing with diamond and graphite nearly always include one or more questions asking for links between structure and properties. It is important that you can explain those links.

Now test yourself

TESTED

6 An individual ionic bond is weak. Why do ionic compounds have high melting points?
7 How is a covalent bond formed?
8 Why are most covalent compounds gas or liquid at room temperature?
9 Why is graphite a good lubricant?
10 Why is diamond so hard?

Answers online

Modern materials

Fullerenes

REVISED

The fullerenes are a group of allotropes of carbon. They are made of balls, 'cages' or tubes of carbon atoms.

Allotrope: Different physical forms of the same substance

Nanotubes

REVISED

Carbon nanotubes are one type of fullerene. They are molecular scale tubes of graphite-like carbon with remarkable properties.
- Carbon nanotubes are among the stiffest and strongest fibres known to mankind.
- They have amazing electrical properties: depending on their exact structure they can have higher electrical conductivity than copper – all in a tube about 10 000 times thinner than a human hair.

Carbon nanotubes are formed when graphite layers form and then roll up into tubes rather than being deposited in layers (Figure 18.11). The covalently bonded hexagonal carbon sheets make carbon nanotubes incredibly strong and the free electrons give them a high electrical conductivity.

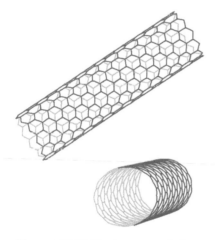

Figure 18.11 Carbon nanotubes.

Other fullerenes

REVISED

Other types of fullerene have different shapes to nanotubes – either balls or 'cages'. One example, Buckminsterfullerene (sometimes known as 'buckyballs') is shown in Figure 18.12. It consists of a giant molecule of 60 carbon atoms in the shape of a ball. The balls can fit together to form a transparent yellow solid called fullerite. Other buckyballs have been made with 70, 76 and 84 carbon atoms. The properties are generally similar to nanotubes and once again they are very strong, but because they have a closed structure they can entrap other molecules so they can have additional applications such as carrying drugs for delivery to specific sites in the body, or in cosmetics, as they can absorb harmful free radicals.

Figure 18.12 Structure of Buckminsterfullerene.

Exam tip

You need to know about the structure and properties of fullerenes in general, and nanotubes in particular. The Buckminsterfullerene mentioned here is just used as an example – you will not be tested directly on that material.

Graphene

REVISED

Graphene is similar to one layer of a graphite molecule. It consists of a single layer of carbon atoms that are bonded together in a hexagonal honeycomb lattice. It has many extraordinary properties.

● It is the thinnest material known to man (only one atom thick).
● It is the lightest material known (1 square metre of graphene weighs around 0.77 milligrams).
● It is 100 to 300 times stronger than steel, making it the strongest material discovered.
● It is the best conductor of heat at room temperature and the best conductor of electricity known.

Nano-particles

REVISED

Silver

We have seen with diamond, graphite, fullerenes and graphene that large assemblages of atoms can have unique properties because of the arrangement of the atoms rather than the actual atoms concerned. All of those materials consist only of carbon atoms, yet they have quite different properties. Another example of this is **nano-scale** silver particles. A nano-scale particle is 1–100 nanometres (nm) in diameter, and at this size silver has different properties from that of naturally found metallic silver. (1 nm = 1×10^{-9} m, or a billionth of a metre).

Silver ions (Ag^+) kill bacteria in water and in wounds. Doctors and paramedics use wound dressings containing silver sulfadiazine or silver nano-materials to treat external infections. Many products now have tiny particles of silver added to them to increase their ability to kill microorganisms.

All of the above are properties of 'standard' silver particles, but nano-particles sometimes have different properties from their bulk material, because they have a very large surface area for their size compared to normal particles. Nano-scale silver particles have a much greater antibacterial action than normal silver, because their tiny size means that they can get to places normal particles cannot, and they may even be small enough to enter living cells.

(H) Nano-scale particles have a wide variety of potential applications in biomedical, optical and electronic fields, but some people are worried about the wide-scale use of nano-particles, particularly silver, and particularly its use in antibacterial products such as soaps and disinfectants, and in clothing.

Although there is no evidence so far of silver nano-particles damaging people's health or the environment, we cannot be certain that we know all the new properties that silver nano-particles have compared to normal silver. As nano-particles are a new development, we cannot tell yet if they have any long-term effects.

Titanium dioxide

Titanium dioxide is a white solid used in house paint and the coating of some chocolates. Titanium dioxide nano-particles are so small that they do not reflect visible light, so they are invisible. They are being used in sun screens as they can block harmful ultraviolet light but do not appear white on the skin. The particles are so small they are easily absorbed through the skin, and they have to be coated to avoid some sorts of skin irritation. Titanium dioxide is often found in food products (e.g. white icing) and it seems that some titanium dioxide nano-particles will be present, even if they have not been specifically added.

(H) Normal titanium dioxide is harmless, but once again the concern among some people is that although there is no direct evidence that titanium dioxide nano-particles are harmful, we do not know for certain that they are harmless.

Smart materials

Smart materials have properties that change reversibly with a change in their surroundings. Smart pigments, for example, that are used in some paint applications, change their colour in response to changes in their environment.

Thermochromic pigments

REVISED

Thermochromic pigments are materials that form the basis of special paints that change their colour at a specific temperature Most thermochromic materials are based on liquid crystal technology, similar to the materials used in flat screen televisions. At specific temperatures, the liquid crystals re-orientate themselves to produce an apparent change in colour. When the temperature drops, they re-orientate back into their original position and colour. Examples of the use of thermochromic pigments include mugs that change colour when they have hot liquids in them, battery power indicators and T-shirts that change colour depending on body temperature.

Photochromic pigments

REVISED

Photochromic pigments change colour with light intensity. Photochromic pigments contain special organic molecules that, when exposed to light, particularly ultraviolet light, change colour. The light breaks a bond in the molecule, which then rearranges itself, forming a molecule with a different colour. When the light source is removed, the molecule returns to its original form.

Photochromic pigments are used in T-shirt design, but perhaps the most wide-scale application is for photochromic lenses in glasses. These allow prescription glasses to function as sunglasses as well, because they darken in bright sunlight.

Shape-memory polymers

REVISED

Shape-memory polymers are 'plastics' that can regain their shape when they are heated. When first heated, the polymer softens, and it can be stretched or deformed – or pressed into a particular shape by a machine. On cooling, it remains in the deformed, changed-shape state. On being reheated, it 'remembers' its original shape, to which it returns. This property is called shape retention. Shape-memory polymers are used in the building industry for sealing around window frames and for the manufacture of sportswear such as helmets and gum-shields. Potential applications are plastic car bodies from which a dent could be removed by heating, and medical sutures ('stitches') that will automatically adjust to the correct tension and be biodegradable, and so will not need to be surgically removed.

Shape-memory alloys

REVISED

Shape-memory alloys are metal alloys that also regain their original shape when they are heated (like shape-memory polymers). Some alloys, in particular some nickel/titanium alloys (often called NiTi or nitinol) and copper/aluminium/nickel alloys, have two remarkable properties:

- pseudoelasticity (they appear to be elastic)
- shape retention (when deformed, they return to their original shape after heating).

Applications include: deformable spectacle frames; surgical plates for joining bone fractures (as the body warms the plates, they put more tension on the bone fracture than conventional plates); surgical wires that replace tendons; and thermostats for electrical devices such as coffee pots.

Polymer gels

REVISED

Hydrogels are polymer gels that absorb or expel water and swell or shrink (up to 1000 times their volume) due to changes in pH or temperature. Hydrogels are cross-linked polymers that, due to the open nature of the cross-linked structure, enable water (or some aqueous solutions) to be absorbed within the structure causing the structure to swell. Small changes in the stimulus (either temperature or pH) control the amount of swelling or shrinking.

Applications include: artificial muscles; granules added to house plant compost to retain water; filling for nappies to allow them to be used for longer without changing and use in houses threatened by forest fires where hydrogels can be more effective than fire-fighting foam.

> **Exam tip**
>
> The specification states that you need to know uses of smart materials, but does not list specific ones. You do not need to learn all the uses listed here, but you should certainly know one example for each material.

Now test yourself

TESTED

11 Fullerenes can be used to deliver drugs inside the body. Why are 'buckyballs' used for this, rather than nanotubes?
12 Silver is a well-known element. Why can we not be sure about the properties of silver nano-particles?
13 What is a thermochromic pigment?
14 What is the difference between a shape-memory polymer and a shape-memory alloy?
15 Why do people add polymer gel granules to the soil around their house plants?

Answers online

Summary

- Metals, ionic compounds, simple molecular covalent substances and giant covalent substances are groups of materials and each group shares certain properties.
- The physical properties of metals can be explained using the 'sea' of electrons/lattice of positive ions structural model.
- Ionic bonding involves the loss or gain of electrons. The resultant ions are held together by electrostatic forces.
- The structure of giant ionic substances explains their physical properties.
- Covalent bonds are formed when atoms share electrons.
- The weak intermolecular forces in simple covalent substances explains their low melting and boiling points.
- The properties of diamond, graphite, fullerenes, carbon nanotubes and graphene can be explained in terms of their structure and bonding.

- Bulk materials do not have the same properties as individual atoms, as demonstrated by diamond, graphite, fullerenes, carbon nanotubes and graphene having different properties despite all containing only carbon atoms, and by nano-scale silver particles exhibiting properties not seen in bulk silver.
- Nano-scale particles of silver and titanium dioxide have properties, which make them useful in antiseptics (silver) and cosmetics (titanium dioxide).
- **(H)** There are possible risks associated with the use of nano-scale particles of silver and titanium dioxide.
- The current research into nano-scale particles is likely to lead to new applications.
- 'Smart' materials include thermochromic pigments, photochromic pigments, polymer gels, shape-memory alloys and shape-memory polymers. Their smart properties lead to specific uses.

Exam practice

1 The diagram below shows an atom of chlorine. Only the electrons in the outer shell are shown.

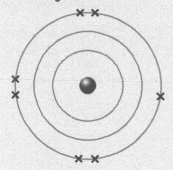

Figure 18.13

a) A chlorine molecule (Cl_2) is made of two chlorine atoms joined by a covalent bond. Draw a diagram to show the arrangement of electrons in a chlorine molecule. [2]

b) Chlorine is gas at room temperature. Explain the link between this and its covalent structure. [2]

c) Chlorine can form ionic bonds with other substances. Explain why ionic bonds could not form between two chlorine atoms. [4]

d) Explain how an ionic bond forms between chlorine and sodium in the formation of sodium chloride. [5]

2 A designer is designing a mug that will change colour when hot water is poured into it. She has a thermochromic pigment that will change colour from red to colourless at the temperature of hot water, but she wants the mug to remain coloured when hot. She also has a blue pigment which is not thermochromic.

a) How could she use the thermochromic pigment to achieve her design? [3]

b) Suggest one other use for thermochromic pigments. [1]

c) Some pigments are called photochromic. What is the difference between a thermochromic pigment and a photochromic pigment? [1]

Answers and quick quiz 18 online

ONLINE

19 Acids, bases and salts

pH

Substances can be classified as acids or alkalis using the pH scale (Figure 19.1). Acids have a pH lower than 7 and alkalis have a pH higher than 7; pH 7 is neutral. pH is measured using a chemical indicator (such as Universal Indicator paper) or an electronic pH meter.

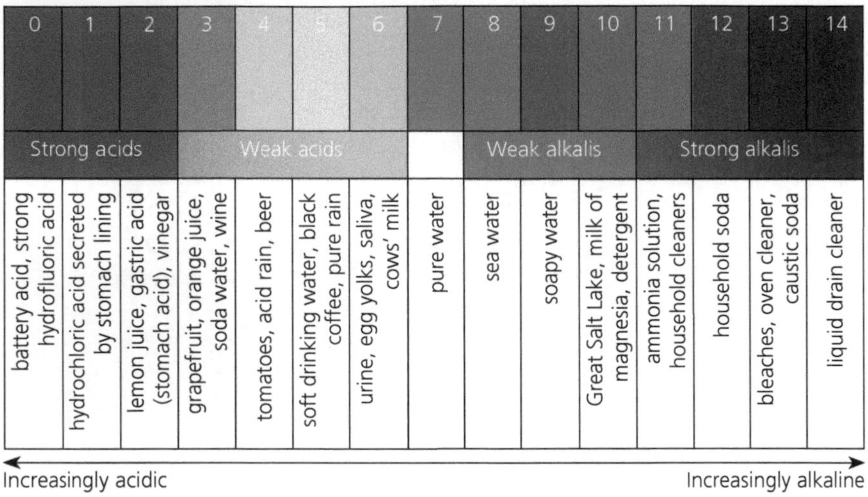

0	1	2	3	4	5	6	7	8	9	10	11	12	13	14
Strong acids			Weak acids					Weak alkalis			Strong alkalis			
battery acid, strong hydrofluoric acid	hydrochloric acid secreted by stomach lining	lemon juice, gastric acid (stomach acid), vinegar	grapefruit, orange juice, soda water, wine	tomatoes, acid rain, beer	soft drinking water, black coffee, pure rain	urine, egg yolks, saliva, cows' milk	pure water	sea water	soapy water	Great Salt Lake, milk of magnesia, detergent	ammonia solution, household cleaners	household soda	bleaches, oven cleaner, caustic soda	liquid drain cleaner

← Increasingly acidic   Increasingly alkaline →

Figure 19.1 The pH scale.

Reactions of metals with acids

Some metals will react with acids. Their reaction depends upon their position in the reactivity series (Figure 19.2).

Metals at the top of the reactivity series react very vigorously and exothermically (occasionally explosively) with dilute acids, whereas those metals at the bottom of the reactivity series do not react at all.

Metals react with acids forming salts and hydrogen.

metal + acid → metal salt + hydrogen

Reactions with hydrochloric acid   REVISED ☐

Reactions involving hydrochloric acid form salts called chlorides.

metal + hydrochloric acid → metal chloride + hydrogen

For example:

magnesium + hydrochloric acid → magnesium chloride + hydrogen

$Mg(s)$ + $2HCl(aq)$ → $MgCl_2(aq)$ + $H_2(g)$

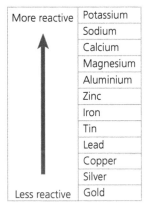

More reactive	Potassium
	Sodium
	Calcium
	Magnesium
	Aluminium
	Zinc
	Iron
	Tin
	Lead
	Copper
	Silver
Less reactive	Gold

Figure 19.2 A reactivity series.

Reactions with sulfuric acid

REVISED

Reactions involving sulfuric acid form sulfates.

metal + sulfuric acid → metal sulfate + hydrogen

For example:

magnesium + sulfuric acid → magnesium sulfate + hydrogen

$Mg(s)$ + $H_2SO_4(aq)$ → $MgSO_4(aq)$ + $H_2(g)$

> **Exam tip**
>
> The charge on the ions is needed to work out the numbers of each ion in the compound. The charges need to balance – equal number of positive and negative charges.

Reactions with nitric acid

REVISED

Reactions involving nitric acid form nitrates.

metal + nitric acid → metal nitrate + hydrogen

For example:

magnesium + nitric acid → magnesium nitrate + hydrogen

$Mg(s)$ + $2HNO_3(aq)$ → $Mg(NO_3)_2(aq)$ + $H_2(g)$

Now test yourself

TESTED

1 A small piece of magnesium ribbon was placed into excess dilute hydrochloric acid in a boiling tube. The temperature of the reaction mixture was recorded using a temperature sensor and displayed on a computer screen.

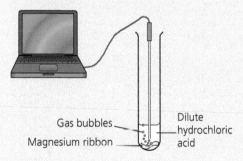

Gas bubbles
Magnesium ribbon
Dilute hydrochloric acid

Figure 19.3

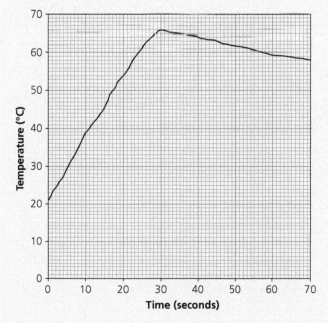

Figure 19.4

a) Use the graph to answer parts i) and ii).
  i)   State the time taken for the reaction to come to an end.
  ii) State the maximum temperature rise recorded during the reaction.
b) Give one advantage of using a temperature sensor and computer to record the temperature.
c) i)   Other than the temperature change, give two observations which suggest that a chemical change is occurring when magnesium is added to dilute acid.
  ii) The following table shows the colours of Universal Indicator at different pH ranges. One of the products of the reaction is magnesium chloride. Magnesium chloride solution is neutral. Give the colour of Universal Indicator in magnesium chloride solution.

Colour	Red	Orange	Yellow	Green	Blue	Navy blue	Purple
pH range	0–2	3–4	5–6	7	8–9	10–12	13–14

2 Use the table of ions below to write word equations and balanced symbol equations for the following reactions between metals and acids:
  a) magnesium and nitric acid
  b) lithium and hydrochloric acid
  c) calcium and sulfuric acid.

Positive ions		Negative ions	
Lithium	Li^+	Chloride	Cl^-
Calcium	Ca^{2+}	Sulfate	SO_4^{2-}
Magnesium	Mg^{2+}	Nitrate	NO_3^-

Answers online

Neutralisation reactions

- Acids react with metal oxides, metal hydroxides (both called bases) and metal carbonates.
- Soluble bases are called alkalis, such as sodium hydroxide.
- These reactions are all exothermic; they give out heat to their surroundings.
- When acids react with bases/alkalis this is called neutralisation. For example:
  - acid + base → metal salt + water
    hydrochloric acid + calcium oxide → calcium chloride + water
    $2HCl(aq)$ + $CaO(s)$ → $CaCl_2(aq)$ + $H_2O(l)$
  - acid + alkali → metal salt + water
    nitric acid + sodium hydroxide → sodium nitrate + water
    $HNO_3(aq)$ + $NaOH(aq)$ → $NaNO_3(aq)$ + $H_2O(l)$
- Neutralisation reactions involving acids and alkalis can be characterised by the ionic equation:
  $H^+(aq) + OH^-(aq) → H_2O(l)$
- Metal carbonates react with acids forming metal salts, water and carbon dioxide gas. For example:
  - acid + metal carbonate → metal salt + water + carbon dioxide
    sulfuric acid + magnesium → magnesium + water + carbon dioxide
                   carbonate   sulfate
    $H_2SO_4(aq)$ + $MgCO_3(s)$ → $MgSO_4(aq)$ + $H_2O(l)$ + $CO_2(g)$
- The chemical test for carbon dioxide gas is to pass it through limewater (calcium hydroxide solution). If the gas is carbon dioxide then the limewater turns 'cloudy' or 'milky'.

- The test for a carbonate (and the carbonate ion, CO_3^{2-}) is to add an acid. If the substance effervesces (gives off bubbles), producing carbon dioxide gas, then it is a carbonate (or it contains the carbonate ion). In a similar way, this reaction can be used as a test for acidic substances.

Titrations

A titration is a method of preparing solutions of soluble salts and a way of determining the relative and actual concentrations of solutions of acids or alkalis. During a titration an indicator such as phenolphthalein is added to a known volume of alkali, usually put inside a conical flask as shown in Figure 19.5.

The acid is usually put into a burette and the concentration of either the acid or the alkali is known. The acid is then carefully added to the alkali until the indicator just turns colour (phenolphthalein is pink in an alkali and colourless in an acid). The volume of acid needed to just neutralise the alkali is recorded. The unknown concentration can then be determined using:

$$\text{concentration of acid} \times \text{volume of acid} = \text{concentration of alkali} \times \text{volume of alkali}$$

Testing for a sulfate

The standard chemical test for sulfate ions (SO_4^{2-}) involves reacting a solution of the sulfate with a solution of barium chloride. A white precipitate of insoluble barium sulfate is formed. For example, sodium sulfate reacts with barium chloride forming sodium chloride (which is soluble) and a precipitate of barium sulfate:

sodium sulfate + barium chloride → sodium chloride + barium sulfate

$Na_2SO_4(aq) + BaCl_2(aq) \rightarrow 2NaCl(aq) + BaSO_4(s)$

$Ba^{2+}(aq) + SO_4^{2-}(aq) \rightarrow BaSO_4(s)$

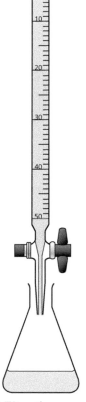

Figure 19.5 Titration apparatus.

Now test yourself

TESTED ☐

3 Write ionic equations for the reactions between the following chemicals:
   a) lithium sulfate and barium chloride
   b) potassium hydroxide and hydrochloric acid
   c) lithium hydroxide and nitric acid
   d) sodium hydroxide and sulfuric acid.

Answers online

Preparing soluble salts

Soluble salts such as magnesium sulfate, can be prepared by reacting an acid, such as sulfuric acid, with a metal base such as magnesium oxide. As the resultant salt (such as magnesium sulfate) is soluble, it cannot be separated using filtration and an evaporation technique must be used, as shown in Figure 19.6.

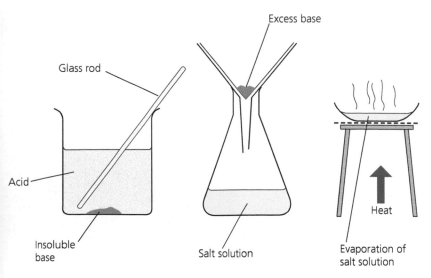

Figure 19.6 Solution of a soluble salt prepared by neutralising acid.

The excess insoluble base is first separated from the soluble salt using filtration before the water is evaporated off using a heat source and an evaporating basin. Small crystals form inside the basin as the water evaporates.

Now test yourself

4 The diagram below shows the stages in making the compound copper sulfate by reacting copper carbonate with dilute sulfuric acid.

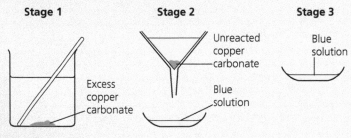

Figure 19.7

a) State why copper carbonate is added in excess.
b) Choose, from the list below, the name for the process occurring in:
   i) stage 2
   ii) stage 3.
   boiling   evaporating   dissolving   filtering
c) The reaction that takes place in stage 1 can be described by the following word equation.
   sulfuric acid + copper carbonate → copper sulfate + water + carbon dioxide
   i) Choose, from the list below, the name of the group of compounds to which copper sulfate belongs.
   acid   base   salt
   ii) All of the substances in the above equation are compounds. State how compounds are different from elements.
d) If sodium carbonate were used instead of copper carbonate, give the chemical name of the crystals formed in the evaporating basin in stage 3.

Answers online

Identifying chemicals

The reactions of an unknown chemical with acids or alkalis and other (known) chemicals can be used as a way of identifying the chemical. This is a common question in the examination. The reactions may be presented in the form of a flow chart and you may have to identify the unknown chemical, or you could be given the name of a substance and asked to work out what chemicals react with it in ways that are given in the diagram.

Exam tip

In the exam, study the flow chart carefully and use a pencil to write your ideas in and around the diagram – you can always cross or rub them out if you change your mind.

Example

Figure 19.8 shows some reactions of dilute hydrochloric acid.

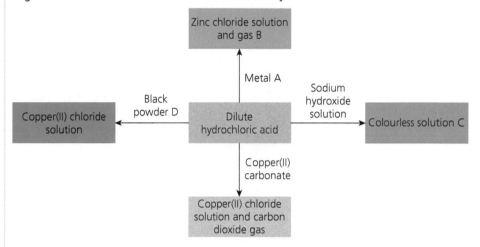

Figure 19.8 **A flowchart showing some reactions of dilute hydrochloric acid.**

Give the name for:
1. metal A
2. gas B
3. colourless solution C
4. black powder D.

Answers

1. Zinc, because zinc chloride is formed when it is reacted with hydrochloric acid.
2. Hydrogen, because reactions of metals with acids produce hydrogen gas.
3. Sodium chloride solution, because the reaction of hydrochloric acid with sodium hydroxide produces the salt sodium chloride.
4. Copper oxide, because it reacts with hydrochloric acid producing copper(II) chloride.

Exam tip

The style of question in the example is one commonly used in the exam. You are given a diagram and you have to identify different components from the clues. Practise as many of these questions as you can from the past papers.

Summary

- Substances can be classified as acidic, alkaline or neutral in terms of the pH scale.
- Solutions of acids contain hydrogen ions and alkalis contain hydroxide ions.
- Neutralisation occurs between the reaction of hydrogen ions with hydroxide ions to form water: $H^+(aq) + OH^-(aq) \rightarrow H_2O(l)$
- Acids react with some metals. The extent of the reaction depends on the metal's position in the reactivity series.
- The reaction of dilute acids with bases (and alkalis) is called neutralisation, and these reactions are exothermic (give out heat).
- A titration is a method to prepare solutions of soluble salts and can be used to determine the relative and actual concentrations of solutions of acids or alkalis.
- Reactions involving hydrochloric acid form chlorides, reactions involving nitric acid form nitrates and reactions involving sulfuric acid form sulfates.
- The test for a sulfate is to react it with barium chloride – an insoluble white precipitate of barium sulfate is formed.
- The reaction of dilute acids and carbonates is also exothermic; carbonates effervesce in acid, giving out carbon dioxide gas.
- The test to identify carbon dioxide gas is to pass the gas through limewater. If it turns milky then the gas is carbon dioxide.
- Soluble salts, such as copper sulfate, can be made from the reaction of insoluble bases and carbonates with acids. These soluble salts can form crystals.
- Word equations and balanced symbol equations can be used to describe the reactions of metals, bases (including alkalis) and carbonates with hydrochloric acid, nitric acid and sulfuric acid.

Exam practice

1 The chemicals and apparatus needed to prepare crystals of copper(II) sulfate are shown below.

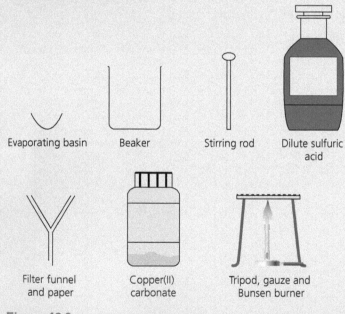

Figure 19.9

There are three stages to the preparation of copper(II) sulfate crystals.

a) Using the chemicals and apparatus shown above, draw diagrams to show how each stage would be carried out: [3]
 Stage 1: Reacting copper(II) carbonate with dilute sulfuric acid until no more dissolves
 Stage 2: Removing unreacted copper(II) carbonate
 Stage 3: Obtaining crystals of copper(II) sulfate

b) Name the process described in stage 2. [1]

c) Choose substances from the box to copy and complete the word equation for the reaction taking place. [1]

| carbon dioxide | hydrogen | copper(II) sulfate | water | copper(II) chloride |

copper(II) carbonate + sulfuric acid → _____ + _____ + _____

(From WJEC GCSE Chemistry 1 Foundation Jun 16 Q1)

2 The apparatus below can be used to measure the temperature as a neutralisation reaction takes place.

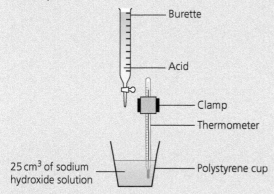

Figure 19.10

The graph shows how the temperature changes when acids A and B are added separately to 25 cm³ of sodium hydroxide solution.

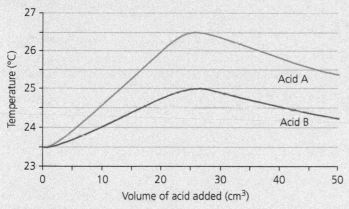

Figure 19.11

a) Use the graph to find:
  i) the volume of acid required to neutralise the sodium hydroxide solution in both experiments [1]
  ii) the maximum temperature rise for acid B. [1]
b) State which acid, A or B, is stronger and give a reason for your answer. [1]
c) Describe how an indicator could be used to find the exact volume of acid needed for neutralisation. [3]

(From WJEC GCSE Chemistry Unit 2 Foundation Tier Sample Assessment Materials Q7)

3 a) A student carries out a series of chemical tests on three unknown solutions, A, B and C. Her results are recorded in the table below. Use all the information to identify reagents X and Y and solutions A and B. [4]

	Add dilute HCl	Add BaCl₂(aq)	Add reagent X	Add reagent Y
A	No reaction	White precipitate forms	Pale green precipitate forms	No reaction
B	Fizzes	No reaction	Pungent smell given off	White precipitate forms
C	No reaction	No reaction	No reaction	Yellow precipitate forms

b) Give the balanced **symbol** equation for the reaction that takes place between sodium carbonate and dilute nitric acid. [2]

c) The equation below represents the reaction occurring between copper(II) chloride solution and sodium hydroxide solution.

$CuCl_2 + 2NaOH \rightarrow Cu(OH)_2 + 2NaCl$

Write the ionic equation for this reaction. Include state symbols. [2]

(From WJEC GCSE Chemistry Unit 2 Higher Tier Sample Assessment Materials Q7)

4 a) The diagram below shows some reactions of dilute sulfuric acid.

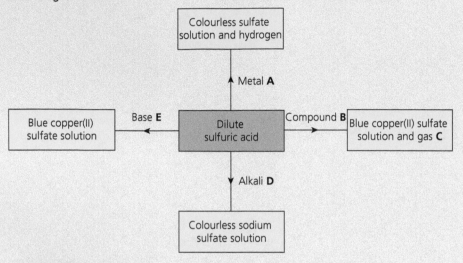

Figure 19.12

Give the names of each of the substances **A** to **E**. [5]

b) Give the chemical formula of ammonium sulfate. [1]

(From WJEC GCSE Chemistry 1 Higher Jun 16 Q5)

5 A class of students was asked to carry out a neutralisation reaction as part of an experiment to prepare crystals of a salt. They carried out the first stage of the experiment using the apparatus shown below.

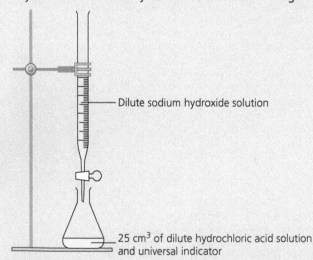

Figure 19.13

This stage of the experiment was carried out three times by five different groups. Their results are shown below.

Group	Volume of sodium hydroxide needed to neutralise the hydrochloric acid (cm³)		
1	24.2	24.8	24.7
2	24.6	24.8	24.7
3	25.1	25.3	25.8

| 4 | 24.5 | 24.5 | 24.5 |
| 5 | 24.9 | 25.0 | 25.1 |

a) Two of the results in the table should not be used when calculating a mean value for each group. Identify these results. [1]

b) Universal indicator is used to identify when neutralisation has occurred. State how the pH will change as the reaction takes place and give the colour of the solution when neutral. [2]

c) In the final stage of the experiment, the students used the following apparatus to crystallise their salt from a solution without universal indicator.

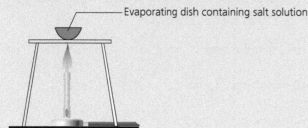

Evaporating dish containing salt solution

Figure 19.14

i) Name the colourless liquid removed during evaporation. [1]

ii) Give the chemical name for the salt formed. [1]

d) The experiment was repeated using sulfuric acid. Complete the word equation for the reaction. [1]

sodium hydroxide + sulfuric acid → _____ + _____

(From WJEC Chemistry 1 Foundation Jan 16 Q4)

Answers and quick quiz 19 online

ONLINE ☐

20 Metals and their extraction

Metals, ores and reactivity

- An ore is a substance found in the Earth's crust containing metal atoms combined with other elements. Examples are haematite and magnetite (iron), and bauxite (aluminium).
- Metals can be extracted from their ores by chemical reactions or electrolysis.
- Some very unreactive metals, like gold, silver and platinum, can be found uncombined with other elements.
- The difficulty of extracting metals from their ores (amount of energy needed) increases as their reactivity increases.
- Metals can be arranged in order of their reactivity in a reactivity series (Figure 20.1).

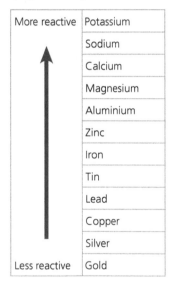

More reactive	Potassium
	Sodium
	Calcium
	Magnesium
	Aluminium
	Zinc
	Iron
	Tin
	Lead
	Copper
	Silver
Less reactive	Gold

Figure 20.1 A reactivity series.

- Any metal can displace a metal lower than it in the reactivity series from a solution of one of its salts (called a displacement reaction) (e.g. magnesium can displace copper from a solution of copper sulfate, forming magnesium sulfate).

magnesium + copper sulfate → copper + magnesium sulfate

$Mg(s)$ + $CuSO_4(aq)$ → $Cu(s)$ + $MgSO_4(aq)$

- The relative reactivities of metals can be demonstrated by competition reactions, such as the thermite reaction, where a mixture of powdered aluminium and iron(III) oxide is ignited by a high-temperature fuse and molten iron is formed. This reaction is used in the rail industry to weld rails together on a track.

aluminium + iron(III) oxide → aluminium oxide + iron

$Al(s)$ + $Fe_2O_3(s)$ → $Al_2O_3(s)$ + $Fe(s)$

Now test yourself

TESTED

1 This question is about the reactivity of metals.
   a) A teacher carried out the following two experiments in a fume cupboard.
   Experiment 1: A mixture of aluminium powder and iron oxide was heated strongly using the equipment shown in Figure 20.2. The reaction that took place can be summarised by the following word equation:

   aluminium + iron oxide → aluminium oxide + iron

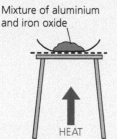

Mixture of aluminium and iron oxide

HEAT

Figure 20.2 **Experiment 1.**

Experiment 2: The experiment was then repeated using a mixture of iron powder and copper oxide. The word equation for this reaction is shown below:

iron + copper oxide → iron oxide + copper

   i) Use the results of the two reactions to place the three metals, aluminium, copper and iron, in order of decreasing reactivity.
   ii) The teacher said that iron oxide, in Experiment 1, and copper oxide, in Experiment 2, had both been 'reduced'. State the meaning of the term reduced.

   b) Zinc is more reactive than copper. Excess zinc powder was added to the blue copper sulfate solution. During the reaction, the blue solution became colourless and a brown solid was formed, as in Figure 20.3.

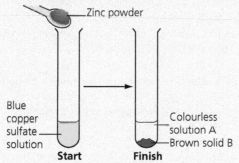

Zinc powder

Blue copper sulfate solution

Start

Colourless solution A
Brown solid B

Finish

Figure 20.3 **Zinc powder reacting with copper sulfate solution.**

   i) Name the colourless solution, A.
   ii) Name the brown solid, B.

2 Metal X is suspected to lie between magnesium and iron in the reactivity series. Describe and explain how you would show this was true using the following chemicals: magnesium ribbon, iron filings and metal X, solutions of magnesium nitrate, iron nitrate and the nitrate of metal X.

Answers online

Extracting metals from their ores

Many chemical reactions used to extract metals from their ores involve the processes of reduction and oxidation:

- Reduction involves the removal of oxygen atoms from a chemical OR the gain of electrons.
- Oxidation involves the addition of oxygen atoms to a chemical OR the loss of electrons.

    OiLRiG – Oxidation is the Loss of electrons; Reduction is the Gain of electrons.

Extraction of iron

The extraction of iron from iron ore is an example of a reduction / oxidation reaction. In this reaction, iron oxide ore is reduced to iron at high temperatures (in a blast furnace) by the oxidation of carbon (from coke), while carbon and carbon monoxide are oxidised to carbon dioxide.

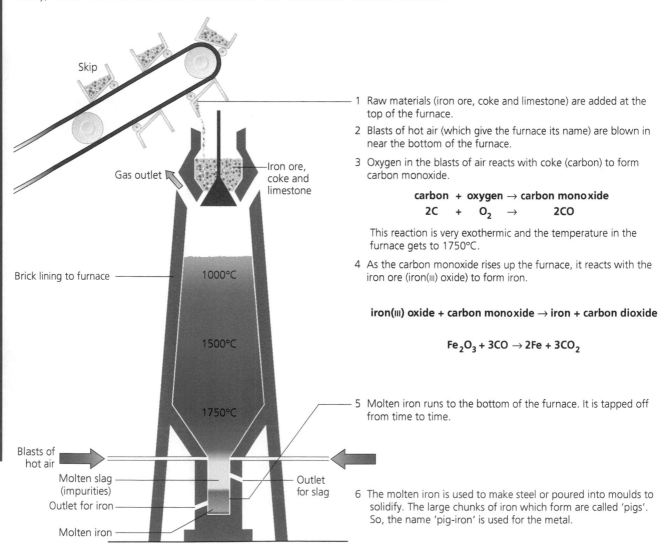

1 Raw materials (iron ore, coke and limestone) are added at the top of the furnace.

2 Blasts of hot air (which give the furnace its name) are blown in near the bottom of the furnace.

3 Oxygen in the blasts of air reacts with coke (carbon) to form carbon monoxide.

$$\text{carbon} + \text{oxygen} \rightarrow \text{carbon monoxide}$$
$$2C + O_2 \rightarrow 2CO$$

This reaction is very exothermic and the temperature in the furnace gets to 1750°C.

4 As the carbon monoxide rises up the furnace, it reacts with the iron ore (iron(III) oxide) to form iron.

$$\text{iron(III) oxide} + \text{carbon monoxide} \rightarrow \text{iron} + \text{carbon dioxide}$$

$$Fe_2O_3 + 3CO \rightarrow 2Fe + 3CO_2$$

5 Molten iron runs to the bottom of the furnace. It is tapped off from time to time.

6 The molten iron is used to make steel or poured into moulds to solidify. The large chunks of iron which form are called 'pigs'. So, the name 'pig-iron' is used for the metal.

Figure 20.4 A blast furnace.

Extraction of aluminium

- Aluminium is higher than iron and carbon in the reactivity series, so we need much more energy to extract it from its ore and reduction / oxidation reactions involving carbon cannot be used.

- Aluminium (and higher metals in the reactivity series) are extracted using electrolysis.
- Aluminium oxide is heated to high temperatures making it melt. This allows positively charged aluminium ions to move towards the negatively charged cathode, forming aluminium atoms that can be extracted as bulk aluminium metal.

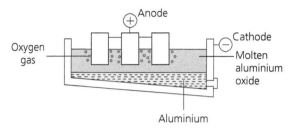

Figure 20.5 Electrolysis of aluminium oxide.

Sustainability issues with the extraction of metals from their ores

- Siting of the plant (large land use, contamination of land), environmental impact, effect on local population.
- Fuel and energy costs.
- Greenhouse gas emissions.
- Impact of recycling.
- Effects of the extraction of the ores on the environment and the local population.

Transition metals

The transition metals are the metallic elements found in the centre of the Periodic Table.

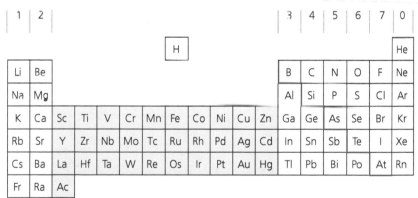

Figure 20.6 The transition metals (shown in blue).

Transition metals have the following physical and chemical properties:
- malleable – can be bent or hammered into shapes easily.
- high melting points (with the exception of mercury, which is liquid at room temperature).
- good conductors of electricity and heat.
- form ions with different charges, for example, copper can lose one electron and form Cu^+ ions or it can lose two electrons and form Cu^{2+} ions.
- less chemically reactive than the alkali metals such as sodium and potassium.
- form coloured chemical compounds.

Electrolysis

Electrolysis is a chemical reaction involving the splitting up of a compound by an electric current passing through a conducting liquid. The conducting liquid is called the electrolyte and contains positively charged ions called cations, and negatively charged ions called anions. The current enters the electrolyte via two solid conductors called electrodes. The positive electrode is called the anode, and the negative electrode is called the cathode. The current is carried through the electrolyte by the movement of the ions. The negative anions move towards the anode. The positive cations are attracted to the cathode.

Electrolysis is used in electroplating, the purification of copper and the manufacture of sodium hydroxide (and hydrogen gas and chlorine gas).

Electrolysis of molten salts

The simplest case of electrolysis is one where the electrolyte contains only two ions, such as the electrolysis of molten zinc chloride, $ZnCl_2$, shown in Figure 20.7.

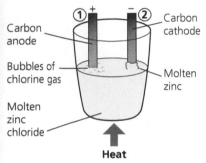

Carbon anode
Carbon cathode
Bubbles of chlorine gas
Molten zinc
Molten zinc chloride
Heat

① Chloride ions change into chlorine gas as they give up electrons to the anode
$2Cl^-(l) \rightarrow Cl_2(g) + 2e^-$

② Zinc ions are changed into zinc atoms as they gain electrons from the cathode
$Zn^{2+}(l) + 2e^- \rightarrow Zn(s)$

Figure 20.7 The electrolysis of molten zinc chloride.

Properties and uses of metals

Iron (steel)

● **Properties**: Can be magnetised, good conductor of heat and electricity. Low-carbon steel is: tough, ductile, malleable, strong but has poor resistance to corrosion. High-carbon steel is: hard, very strong, wear resistant but it is more brittle than low-carbon steel. Stainless steel is resistant to corrosion.
● **Uses**: Magnets, transformer cores, car body panels, cutting tools, ball bearings, railway lines, cutlery, kitchen equipment.

Aluminium

● **Properties**: Strong, low density, good conductor of heat and electricity, resistant to corrosion.
● **Uses**: High-voltage power lines, saucepans, window and greenhouse frames, drinks cans, aeroplane and car parts.

Copper

- **Properties**: Very good conductor of heat and electricity, malleable, ductile, lustrous, attractive colour.
- **Uses**: In alloys like brass and bronze, water pipes, electrical wires, jewellery and ornaments, saucepan bottoms.

Titanium

- **Properties**: Hard, strong, low density, high melting point, resistant to corrosion.
- **Uses**: Jet engine and spacecraft parts, industrial machine parts, car parts, medical implants, strengthening steel, jewellery, sports equipment.

Alloys

Alloys are mixtures of two or more different metals (or carbon) made by combining the molten metals. The properties of the alloy can be modified by varying the types and amounts of the metals in the mixture. Examples of common alloys are brass, bronze and stainless steel (Table 20.1).

Table 20.1 Examples of common alloys.

Alloy	Composition	Examples of uses
Brass	Copper and zinc	Decorative metal parts
		Low friction metal parts (locks, gears), plumbing and electrical applications, musical instruments
Bronze	Copper and tin	Boat and ship fittings, sculptures and statues, guitar and piano strings
		Non-sparking hammers and other tools used in explosive atmospheres
Stainless steel	Iron, carbon and chromium	Cookware and cutlery, surgical instruments
		Car and spacecraft parts
		Large building and bridge construction, jewellery and watches

Now test yourself

3 The following table shows some examples of different types of steel. **Use the information in the table** to answer the questions that follow.

Name	Composition	Properties
Cast iron	Iron, 2–5% carbon	Hard but brittle, corrodes easily
Mild steel	Iron, 0.1–0.3% carbon	Tough, ductile and malleable, good tensile strength, corrodes easily
High carbon steel	Iron, 0.7–1.5% carbon	Harder than mild steel but more brittle, corrodes easily
Stainless steel	Iron and carbon, 16–26% chromium	Hard and tough, hardwearing, doesn't corrode

a) Name the metallic element that is added to iron to make stainless steel.
b) Suggest a reason why cast iron is more brittle than mild steel.
c) Name the type of steel most suitable for making car bodies.
d) Give the main reason why stainless steel is used to make cutlery.

Answers online

Exam tip

Words in bold in a question are like this for a specific reason. Question 3 says to 'Use the information in the table'. If you use other information, from memory, you won't get the marks.

Summary

- Ores found in the Earth's crust contain metals combined with other elements. These metals can be extracted using chemical reactions.
- Some unreactive metals (e.g. gold) can be found un-combined with other elements.
- The difficulty involved in extracting metals increases as their reactivity increases.
- The relative reactivities of metals can be investigated by performing displacement and competition reactions.
- Reduction is the removal of oxygen (from a metal compound); oxidation is the gain of oxygen (by a metal).
- Iron ore, coke and limestone are the raw materials used in the extraction of iron.
- The word and symbol equations for the reduction of iron(III) oxide by carbon monoxide are:
  carbon monoxide + iron(III) oxide → carbon dioxide + iron
  $3CO(g) + Fe_2O_3(s) → 3CO_2(g) + 2Fe(l)$
- The extraction of aluminium requires greater energy input than the extraction of iron. The method used to extract the most reactive metals (including aluminium) is electrolysis.
- The process of electrolysis of molten ionic compounds can be explained in terms of ion movement and electron gain/loss, using the terms electrode, anode, cathode and electrolyte.
- Aluminium is extracted on an industrial scale using large-scale electrolysis. This can be summarised by the following electrode equations in terms of charge and atoms:

$Al^{3+} + 3e^- → Al$
$2O^{2-} - 4e^- → O_2$

- There are considerable environmental, social and economic issues relating to the extraction and use of metals such as iron and aluminium. These include: siting of plants, fuel and energy costs, greenhouse emissions and recycling.
- The uses of aluminium, copper and titanium can be explained in terms of the following relevant properties:
  - steel (iron) – very strong, good conductor of heat and electricity, can be magnetised
  - aluminium – strong, low density, good conductor of heat and electricity, resistant to corrosion
  - copper – very good conductor of heat and electricity, malleable and ductile, attractive colour and lustre
  - titanium – hard, strong, low density, resistant to corrosion, high melting point.
- An alloy is a mixture made by mixing molten metals. Its properties can be modified by changing its composition.
- Reduction can also be explained in terms of the gain of electrons and oxidation in terms of the loss of electrons.
- Transition metals can form ions with different charges, they are less reactive than the alkali metals, form coloured salts, are malleable, have high melting points and are good conductors of electricity and heat.

Exam practice

1 a) Electrolysis is also used to extract aluminium from molten aluminium oxide. On melting, aluminium oxide releases aluminium ions, Al^{3+}, and oxide ions, O^{2-}.

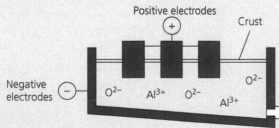

Figure 20.8

   i) State the direction that each of the following ions will move:
      O^{2-} ions: towards _____.
      Al^{3+} ions: towards _____. [1]
   ii) Balance the symbol equation for the overall reaction occurring. [1]
      $2Al_2O_3 →$ ____$Al +$ ____O_2
   iii) Give the **main** reason why this process is expensive. [1]

2 The diagram below shows where materials enter and leave the blast furnace. The labelling is incomplete.

Iron ore, coke and W

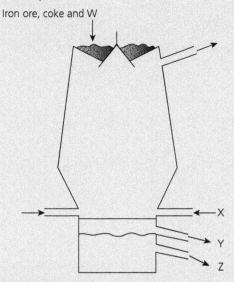

Figure 20.9

| hot air | coke | iron | iron ore | limestone | slag | steel | waste gases |

Choose materials from the box to answer parts (a) and (b).

a) Identify the materials labelled as **W**, **X**, **Y** and **Z** on the diagram. [4]

b) Complete the sentences below.
   i)   The furnace is heated by burning _____ in _____ . [1]
   ii) Impurities are removed by reacting them with _____ to form _____ . [1]

c) Balance the symbol equation that represents the main reaction occurring in the furnace. [1]
$$Fe_2O_3 + \_\_\_CO \rightarrow 2Fe + 3CO_2$$

d) 2000 tonnes of iron ore contains 1100 tonnes of iron. Calculate the percentage of iron in the ore. [2]

(From WJEC GCSE Chemistry 1 Higher Jun 16 Q6)

3 a) A pupil investigated the mass of copper formed when magnesium powder was added to copper(II) sulfate solution.

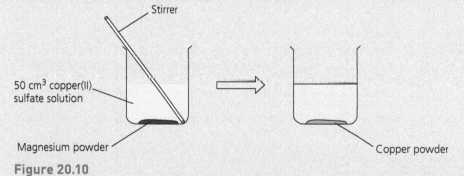

Stirrer

50 cm³ copper(II) sulfate solution

Magnesium powder

Copper powder

Figure 20.10

The results below show the mass of copper formed when different masses of magnesium powder were added to 50 cm³ of copper(II) sulfate solution.

Mass of magnesium added (g)	Mass of copper formed (g)
0	0
0.5	1.3
1.0	2.6
1.5	4.0
2.0	5.3
2.5	6.7

i) On a copy of the grid below plot the results from the table and draw a suitable line. Two points have been plotted for you. [3]

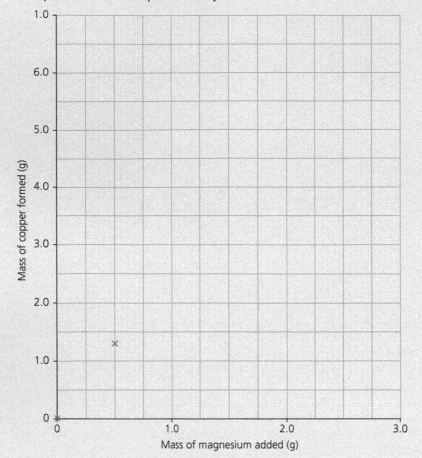

Figure 20.11

ii) Use the graph to find the mass of magnesium needed to form 5.0g of copper. [1]

iii) Copper is obtained by filtering and then washing. State what else must be done to the copper before weighing. [1]

iv) Copy and complete the equation for the reaction taking place by adding the symbols / formulae of the products. [1]

$$Mg + CuSO_4 \rightarrow \underline{\hspace{3cm}} + \underline{\hspace{3cm}}$$

b) When copper is added to a colourless silver nitrate solution, a grey solid and a blue solution are formed. Use this and the information in part (a) to place copper, magnesium and silver in order of reactivity. Explain your reasoning. [2]

(From WJEC GCSE Chemistry 1 Foundation Jun 16 Q6)

4 a) The diagram below shows an electrolysis cell used in the extraction of aluminium.

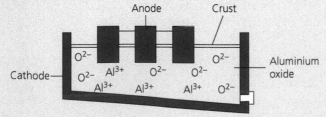

Figure 20.12

i) Give the state (solid, liquid or gas) of the aluminium oxide during this process. [1]

ii) Explain the movement of Al^{3+} and O^{2-} ions during the process. [3]

b) State one property of aluminium that is unusual compared to most other metals. Give a use which relies on this property. [1]

c) Scandium is added to aluminium alloys to increase their strength. The graph In Figure 20.13 shows the relative strength of aluminium alloys, A–D, with and without added scandium. Give the letter of the aluminium alloy where the relative strength is increased by 100 % when scandium is added. Use data from the graph to explain your choice. [2]

(From WJEC GCSE Chemistry 1 Higher Jun 16 Q3)

5 A group of students used the apparatus below to carry out a displacement reaction between zinc powder and copper(II) sulfate solution.

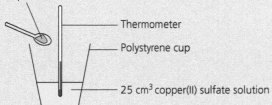

Excess zinc powder

Thermometer

Polystyrene cup

25 cm³ copper(II) sulfate solution

Figure 20.14

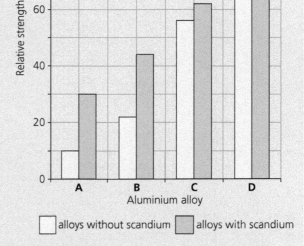

alloys without scandium ▢   alloys with scandium ▨

Figure 20.13

Excess zinc was added to 25 cm³ of the copper(II) sulfate solution at room temperature. The temperature was recorded every 20 s. The results are shown in the table below.

Time after adding the zinc powder to the copper(II) sulfate solution (s)	Temperature of the reaction mixture (°C)		
	Result 1	Result 2	Mean
0	22.0	22.0	22.0
20	22.8	23.0	22.9
40	24.8	25.2	25.0
60	27.3		27.1
80	26.6	26.6	26.6
100	25.7	25.9	25.8
120	24.8	24.4	24.6

a) From the data in the table, calculate the missing result for 60 s that must have been used in working out the mean value. [1]

b) On a copy of the grid below, plot the time after adding the zinc powder against the mean temperature of the reaction mixture and draw a suitable line. [3]

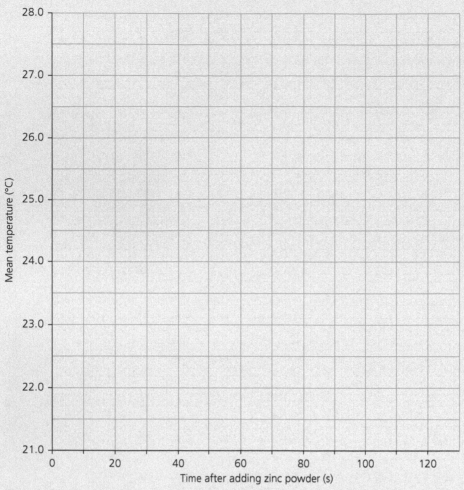

Figure 20.15

c) One of the students checked the thermometer reading 15 minutes later. State what the temperature would be at this point. Give a reason for your answer. [2]

d) Explain why the results recorded in the table can be described as repeatable. [2]

e) The maximum temperature recorded is not as high as expected. Give the main reason for this and suggest one way that this effect could be reduced. [2]

f) Balance the following symbol equation that represents the displacement reaction that takes place between zinc and silver nitrate solution. [1]

$$Zn + \underline{\quad} AgNO_3 \rightarrow Zn(NO_3)_2 + \underline{\quad} Ag$$

(From WJEC GCSE Chemistry 1 Higher Jan 16 Q2)

Answers and quick quiz 20 online

ONLINE

21 Chemical reactions and energy

Making and breaking bonds

Bonds and energy

When a chemical reaction occurs, it is usually accompanied by energy changes. When a reaction happens, energy changes are caused by the need to break bonds and make new ones. Energy is required to break a bond, and released when a bond is formed.

Exothermic and endothermic reactions

When a reaction takes in or gives out energy to the surroundings, it is in the form of heat.
- Reactions that give out energy are called **exothermic reactions**.
- Reactions that take in energy overall are called **endothermic reactions**.

Energy profiles

We can plot the energy changes during a chemical reaction as an **energy profile**. Exothermic and endothermic reactions have different energy profiles (see Figure 21.1).
- In an exothermic reaction, the energy of the products is less than the energy of the reactants, because energy, in the form of heat, has been given out to the surroundings.
- In an endothermic reaction, the energy of the products is greater than that of the reactants, because energy has been taken in from the surroundings.

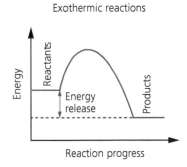
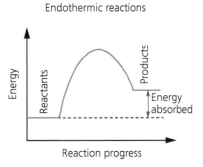

Figure 21.1 Energy changes in endothermic and exothermic reactions.

Activation energy

Look at the energy profiles in Figure 21.1. Before the energy level of the products is reached, in both types of reaction, the level always rises. The rise indicates the energy needed to start the reaction, which is known as the **activation energy**. For two chemicals to react, they must collide in the correct orientation, but also with enough kinetic energy to start a reaction. Such a collision is termed a **successful collision** (most collisions between particles are unsuccessful). The activation energy indicates the level of energy required for successful collisions, and will vary with different reactions (see Figure 21.2).

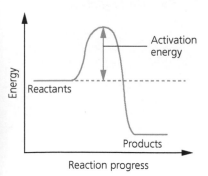

Figure 21.2 The activation energy of a reaction.

Calculating energy changes

The energy changes in a chemical reaction can be calculated in terms of the energy needed to break bonds and that produced in forming bonds.

Combustion is a reaction involving a fuel burning in oxygen. In the case of the complete combustion of hydrocarbons, such as methane, the reactants are the hydrocarbon and oxygen, and the products are carbon dioxide and water. The combustion reaction involves breaking bonds in the reactants and forming bonds to make the products. A hydrocarbon molecule contains only carbon and hydrogen atoms and during the reaction, those atoms form bonds with oxygen atoms.

Consider the combustion of methane (see Figure 21.3).

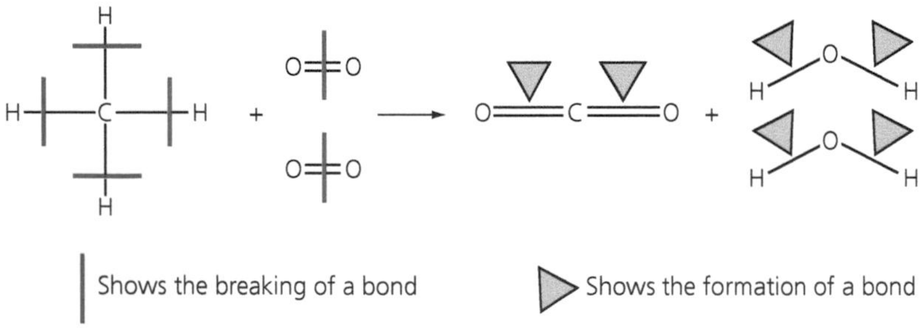

Shows the breaking of a bond   Shows the formation of a bond

Figure 21.3 Combustion of methane involves the breaking and forming of bonds.

$$CH_4(g) \quad + \quad 2O_2(g) \quad \rightarrow \quad CO_2(g) \quad + \quad 2H_2O(g)$$

methane + oxygen → carbon dioxide + water

- The **breaking** of a bond is **endothermic**. This means it requires energy to be put in. In this example, four C-H bonds and two O=O bonds are broken.
- The **formation** of a bond is **exothermic** – it gives out energy. In this example, two C=O bonds and four O-H bonds are formed.
- The difference between the total energy needed to break all the bonds and the total energy given out when the new bonds are formed determines whether the overall reaction is exothermic or endothermic.

Bond energy data

Table 21.1 shows the energy values for some covalent bonds.

Table 21.1 Energy values for covalent bonds.

Bond	Bond energy (kJ)
O=O	496
C–H	412
H–H	436
C=O	743
O–H	463
C–C	348
N≡N	944
C=C	612
N–H	388

> **Exam tip**
>
> You do not need to learn the values for bond energy of different bonds – they will always be given in the question.

Breaking and forming bonds involves the same value for energy change, except that when bonds are broken, the energy is used, and when bonds are formed the energy is released.

> **Worked example 1**
>
> The complete combustion of methane has the equation:
>
> $CH_4(g) + 2O_2(g) \rightarrow CO_2(g) + 2H_2O(g)$
>
> The bonds broken are four C–H bonds and two O=O bonds (see Table 21.1), so the total energy put in is:
>
> $4 \times C{-}H = (4 \times 412) = 1648\,kJ$
>
> $2 \times O{=}O = (2 \times 496) = 992\,kJ$
>
> **Total energy in = 2640 kJ**
>
> The bonds formed are two C=O bonds and four O–H bonds:
>
> $2 \times C{=}O = (2 \times 743) = 1486\,kJ$
>
> $4 \times O{-}H = (4 \times 463) = 1852\,kJ$
>
> **Total energy out = 3338 kJ**
>
> The overall change is total energy in – total energy out = 2640 – 3338 = –698 kJ.
>
> In this example, as is the case with burning all fuels, more energy is given out than is taken in, and so the reaction is exothermic and gives out energy.

Worked example 2

The decomposition of hydrogen bromide has the equation:

$$2HBr(g) \rightarrow H_2(g) + Br_2(g)$$

The bonds broken are two H–Br bonds. The bond energy of an H-Br bond = 366 kJ/mol.

The total energy put in is:

$$2 \times H–Br = (2 \times 366) = 732 \, kJ$$

Total energy in = 732 kJ

The bonds formed are one H–H bond and one Br–Br bond. The bond energy of an H–H bond = 436 kJ/mol; the bond energy of an Br–Br bond = 193 kJ/mol.

The total energy released is:

$$1 \times H–H = 436 \, kJ$$

$$1 \times Br–Br = 193 \, kJ$$

Total energy out = 629 kJ

The overall change is **total energy in – total energy out** = 732 – 629 = +103 kJ.

In this example, more energy is taken in than is given out, and so the reaction is **endothermic** and takes in energy.

Exam tip

Remember when calculating total energy in – total energy out that a positive value means the reaction is endothermic and a negative value indicates an exothermic reaction.

Now test yourself

1 What is an endothermic reaction?
2 Is energy taken in or given out when a chemical bond forms?
3 Exothermic reactions give out energy and so need no energy to start the reaction. True or false?
4 Define the term successful collision.
5 The energy change in a chemical reaction is –345 kJ/mol. Is this reaction exothermic or endothermic?

Answers online

Summary

- Exothermic reactions transfer heat energy to the surroundings.
- In exothermic reactions, the energy of the products is less than that of the reactants.
- In endothermic reactions, the energy of the products is greater than that of the reactants.
- Endothermic reactions absorb heat energy from the surroundings.
- Energy profiles indicate whether a reaction is exothermic or endothermic.
- The activation energy is the energy needed for a reaction to occur.
- Bond energy data can be used to calculate overall energy change for a reaction and to identify whether it is exothermic or endothermic.

Exam practice

1 When hydrocarbon fuels are burnt in a plentiful supply of oxygen, they undergo complete combustion. e.g.

$$C_3H_8 \quad + \quad 5O_2 \quad \rightarrow \quad 3CO_2 \quad + \quad 4H_2O$$

propane   oxygen   carbon dioxide   water

The diagram below shows the structural formulae for each of the chemicals involved.

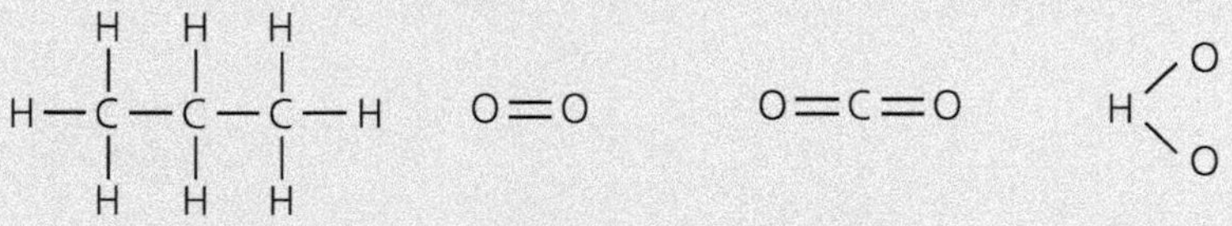

propane   oxygen   carbon   water
dioxide

Figure 21.4

The bond energies for the bonds are:

Bond	Bond energy (kJ/mol)
C–H	413
O=O	496
C=O	734
O–H	463
C–C	348

a) Calculate the energy change in this reaction.   [5]
b) Is this reaction exothermic or endothermic? Give a reason for your answer.   [1]

When there is insufficient oxygen, the hydrocarbon undergoes incomplete combustion:

Propane + oxygen → carbon + carbon monoxide + water

The carbon is apparent as the black powder known as 'soot'. The energy change in this reaction is much smaller than for complete combustion.
c) Give **two** disadvantages of incomplete combustion when using propane as a fuel.   [2]

2 Read the passage below and answer the questions which follow.
Ammonium nitrate is used as a key ingredient of 'cold packs'. Cold packs have a variety of uses, including the treatment of minor injuries to reduce swelling. The cold pack cools the skin, which restricts blood flow and therefore reduces swelling (which is caused by leakage of fluid from blood vessels into the damaged tissue).
Cold packs have an outer pouch that contains solid ammonium nitrate, and an inner pouch which contains water. When the pack is squeezed, the inner pouch splits, causing the ammonium nitrate and the water to mix. The ammonium nitrate dissolves, forming ammonium and nitrate ions.
$$NH_4NO_3(s) \rightarrow NH_4^+(aq) + NO_3^-(aq)$$
This reaction absorbs heat from the surroundings. Approximately 326 joules of energy (in the form of heat) are consumed per gram of ammonium nitrate dissolved in water.
a) Some form of 'hot packs' are re-useable. Would an ammonium nitrate cold pack be re-useable? Give a reason for your answer.   [2]
b) When used to treat injury, what is the main source of the heat which is absorbed by the ammonium nitrate solution?   [1]
c) What term describes a chemical reaction which absorbs heat?   [1]
d) What has happened to the NH_4 part of the ammonium nitrate in order to form NH_4^+?   [1]
e) How much energy in kJ would be consumed when 50 g of ammonium nitrate is added to water?   [2]

Answers and quick quiz 21 online

ONLINE

22 Crude oil, fuels and organic chemistry

Oil and the oil industry

Organic chemistry

REVISED

Organic chemistry is the chemistry of carbon-based compounds. It is particularly important because all life on the planet is carbon based, and so many organic chemicals are necessary for life processes. Carbon-based compounds have many industrial uses, too.

Crude oil

REVISED

Crude oil, along with its associated natural gas, provides a large proportion of the power that drives industry and devices in the home. Most plastics use crude oil as a raw material. Supplies of crude oil are limited, and the burning of fossil fuels created from it are thought to threaten the future of the planet due to global warming.

The crude oil that is extracted from oil wells is a complex mixture of **hydrocarbons**; that is, organic compounds containing only carbon and hydrogen. Crude oil was formed from a process of fossilisation under pressure, mostly of simple marine organisms, over millions of years. That is why oil is called a **fossil fuel**. Oil is vital to the world's economy, because so many useful products can be made from it via the process of fractional distillation.

Fractional distillation is a way of separating out different 'fractions' of the oil. The fractions still consist of a mixture of compounds, but with fewer compounds than in the original oil. The different compounds in crude oil have different boiling points. The boiling point is also the temperature at which the gas will condense into a liquid when it is cooled down. Fractional distillation occurs as follows:

- The oil is heated causing it to boil and turn into a gaseous mixture, and as the vapour rises in the fractionating column, it cools.
- As it cools, the chemicals with higher boiling points will be the first to start to condense.
- At various points, the condensing liquid is extracted from the column, as shown in Figure 22.1.
- The larger the molecule is, the higher its boiling point will be, so the extracted molecules get smaller as you go up the column.

The fractions are not pure chemicals, but a mixture of **alkanes** with broadly similar boiling points.

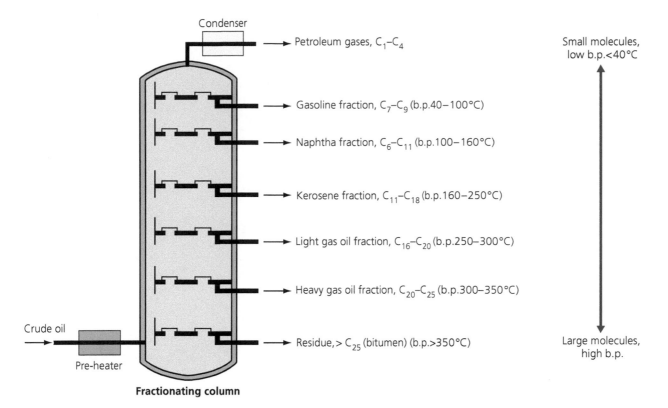

Figure 22.1 The fractional distillation of crude oil.

Chain length, properties and uses

REVISED

The molecules of hydrocarbons are held together by intermolecular forces, which have to be broken for the hydrocarbon to melt or boil. The larger the individual molecules are, the more intermolecular forces there will be, and so more energy is needed to break them – hence the higher melting and boiling points of the longer chain hydrocarbons (Figure 22.2). The increased number of intermolecular forces also makes the longer chain hydrocarbons more **viscous** (i.e. thicker liquids, less easy to pour).

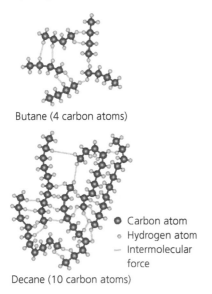

Butane (4 carbon atoms)

Decane (10 carbon atoms)

● Carbon atom
○ Hydrogen atom
— Intermolecular force

Figure 22.2 Intermolecular forces in two hydrocarbons.

The chemicals in the different fractions are used to make a wide variety of useful products, including petrol, diesel, paraffin, methane, lubricating oil and bitumen. Oil products are also used to make plastics.

For a fuel to burn, it must vaporise, as it is the vapour that burns. It is also useful for a fuel to have a low viscosity so that it can be piped from place to place. Lubricating oils, on the other hand, need to be viscous in order to function. The fractions coming off towards the top of the fractionating column therefore make the best fuels, whereas the residue is used for lubricating oils and products like bitumen, which is used in asphalt on roads. The relatively high melting point of bitumen means that the road surface is less likely to melt on hot days.

Oil industry – economic and political issues

REVISED

- The price of oil is set by the oil-producing countries. Other countries cannot do without oil and so have to pay whatever the oil-producing countries demand. This is a greater problem for poorer countries, which are most in need of industrial development.
- The economic success of countries depends upon industry. If war or political upheaval in a major oil-producing country restricts the supplies of oil, this can have a major effect on the world economy.
- If a country depends on another for its supplies of oil, the oil producing country can in effect hold the receiving country to ransom by cutting off the oil supply if there is political disagreement.

Oil industry – social and environmental issues

REVISED

- If a poor country cannot buy all the oil it needs, areas may be deprived of electricity or get an irregular supply (i.e. power cuts).
- The oil industry provides a great many jobs, both directly and indirectly.
- Burning any fossil fuel puts a great deal of carbon dioxide into the environment, and there is general agreement now in the scientific community that this increases the greenhouse effect and contributes to global warming.
- Accidents while oil is being transported around the world by ships have led to major oil pollution incidents which have caused considerable damage to local wild life (and also sometimes to the tourist industry in coastal regions).
- Oil-powered power stations take up a great deal of land and are not particularly attractive, so 'spoiling' areas of countryside for some.

Now test yourself

TESTED

1 What is a hydrocarbon?
2 Why do chemicals with higher boiling points cool towards the bottom of a fractionating column?
3 Why do longer chain hydrocarbons have higher boiling points than shorter chain ones?
4 What is the advantage of a fuel having low viscosity?
5 What is the main advantage of the oil industry to a local population?

Answers online

Fuels

Combustion of fuels

A **combustion reaction** is a chemical reaction where a substance reacts with oxygen, producing heat, light and new products. Any substance that is combustible can theoretically be used as a fuel. Many hydrocarbons are used as fuels, but other types of compound can also be used.

Hydrocarbons burn to form carbon dioxide and water. For example:

$$CH_4 \quad + \quad 2O_2 \quad \rightarrow \quad CO_2 \quad + \quad 2H_2O$$

methane   +   oxygen   →   carbon dioxide   +   water

If the supply of oxygen is limited, then **incomplete combustion** can occur. This results in the formation of carbon (seen as soot), and **carbon monoxide**, as well as carbon dioxide and water. It is important to avoid incomplete combustion of domestic fuels as carbon monoxide is poisonous (and odourless, so people are unaware that they are breathing it in).

Many fuels contain a variety of impurities, the most common of which are compounds of sulfur. Burning results in the formation of **sulfur dioxide**, which can give rise to **acid rain**. Acid rain can result in the death of trees and aquatic animals, and the erosion of limestone buildings and statues.

Measuring the energy released by fuels

The energy released by burning a hydrocarbon fuel can be measured by measuring the temperature rise of a known volume of water when heated by a known mass of fuel. The energy released is calculated using the following equation:

Energy transferred (joules, J) = mass of water heated (grams, g) × 4.2 × temperature rise (°C)

Hydrogen as a fuel

Hydrogen is a combustible gas that has potential use as a fuel. Interest in its use has been sparked by the fact that it only produces water when it burns (no greenhouse gases).

$$2H_2 \quad + \quad O_2 \quad \rightarrow 2H_2O$$

hydrogen + oxygen → water

Hydrogen fuel cells are used in some cars, but at the moment these are uncommon, because there are a number of problems with its use.
- Hydrogen is mainly made by reacting steam with either coal or natural gas. It can also be made by passing electricity through water. The use of coal, gas or electricity (usually produced by burning fossil fuels) means that hydrogen is not a carbon neutral fuel. Although burning hydrogen produces no CO_2, manufacturing hydrogen does.
- Hydrogen is very flammable and can explode, and so is potentially more dangerous than most oil based fuels.
- For storage, hydrogen must be chilled and compressed, and then kept in strong tanks, insulated to keep it cold (liquid hydrogen needs to be kept at around −250°C). Petrol and diesel are much easier to store and transport.

Hydrogen is used as rocket fuel for space exploration. It has specific advantages here − it is extremely light and very powerful, because it burns at over 3000°C.

The fire triangle

The 'fire triangle' is often used as an easily understood symbol to show the factors required for combustion to occur (Figure 22.3). The 'sides' of the triangle are fuel, oxygen and heat. These are needed for a fire to burn so fire-fighting and prevention are based on removing one or more of these factors.

Removing oxygen

Carbon dioxide fire extinguishers and fire blankets deprive the fire of oxygen. Carbon dioxide, being heavier than air, sinks onto the fire and replaces the oxygen around it. Fire blankets seal off the fire and so (once the limited amount of oxygen in the air under the blanket is used up) it goes out. Restricting the supply of oxygen is also the reason for closing doors when evacuating a burning building (to restrict air currents bringing in fresh supplies of oxygen) and for covering a pan containing burning oil with a damp cloth.

Figure 22.3 The fire triangle.

Removing heat

Putting water on a fire removes the heat. It should never be used for electrical fires (because of the danger of shocks) nor for oil fires. As oil and water do not mix, it is ineffective and the water tends to make the oil form droplets, increasing the surface area for contact with oxygen, so making the fire worse.

Removing the fuel

This is primarily used in fire prevention, by the use of non-flammable materials. Clothing can be treated during manufacture to make it fire-resistant and modern buildings often have fire-proof compartments between walls or above ceilings to stop the spread of fire. Managed forests have sections with no trees to prevent fire spreading, and trees may also be felled and removed if there is time to prevent an existing forest fire spreading.

Catalytic cracking

The larger fractions produced in the fractional distillation of crude oil can be converted into more useful smaller molecules by a process called **catalytic cracking**. The heavier fractions are less flammable and more viscous than the lighter ones. This makes them relatively unsuitable as fuels, and although they have other uses they are produced in larger quantities than are needed for those uses. Vapour containing the unwanted fractions is passed over a very hot catalyst surface and this breaks up the long molecules into smaller fragments, which can then be condensed into useful fuels, for example:

$$C_{15}H_{32} \rightarrow 2C_2H_4 + C_3H_6 + C_8H_{18}$$

ethene + propene + octane

The smaller fractions will have shorter chains than the original compounds and therefore have lower boiling points and burn more easily. The hydrocarbons produced are a mixture of alkenes, which contain carbon–carbon double bonds, and alkanes, which do not. Alkenes are useful for making plastics.

Now test yourself

6 Which chemical is essential if a combustion reaction is to take place?
7 The burning of 4 g of ethanol causes 100 cm³ (100 g) of water to rise in temperature by 26°C. How much energy is released per gram of ethanol?
8 State two disadvantages of using hydrogen as a fuel.
9 How do fire blankets work to put out fires?
10 Why are the heavier fractions produced from fractional distillation of crude oil worse as fuels when compared with the lighter fractions?

Answers online

Alkanes and alkenes

Alkanes

REVISED

Alkanes are **saturated** hydrocarbons. They contain single bonds only, and so they are 'saturated' with hydrogen (i.e. they contain as much hydrogen as possible). The difference between different alkanes is the number of carbon atoms they contain. The first four are shown in Table 22.1.

Table 22.1 Alkanes and their structure.

Name of alkane	Chemical/molecular formula	Structural formula
Methane	CH_4	H—C—H with H above and below
Ethane	C_2H_6	H—C—C—H with H above and below
Propane	C_3H_8	H—C—C—C—H with H above and below
Butane	C_4H_{10}	H—C—C—C—C—H with H above and below

> **Exam tip**
>
> You need to know both the structural and molecular formulae for the simple alkanes and alkenes given here, along with the general formulae.

Alkanes are a **homologous series** of chemicals – a series of compounds that have similar properties and the same general formula. The general formula for alkanes is:

$$C_nH_{2n+2}$$

where n is the number of carbon atoms present.

Alkenes

REVISED

Alkenes are unsaturated hydrocarbons, which have a double bond between two carbon atoms. Two alkenes are shown in Figure 22.4. Like alkanes, alkenes are a homologous series. Alkenes have the general formula:

$$C_nH_{2n}$$

Alkenes can be produced from alkanes by **cracking**.

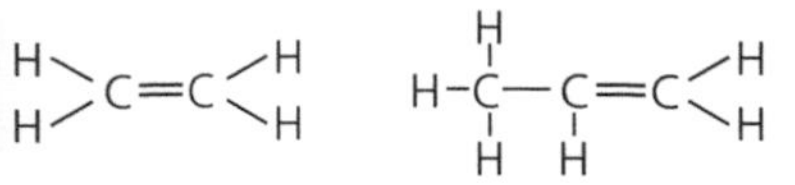

$$H{>}C{=}C{<}{}^H_H \qquad H{-}\underset{H}{\overset{H}{C}}{-}C{=}C{<}{}^H_H$$

Figure 22.4 The structure of two alkenes.

Addition reactions of alkenes

The presence of a carbon–carbon double bond in alkenes means that other atoms can be added to the molecule. Reactions that do this are called **addition reactions**.

When hydrogen is added to an alkene (**hydrogenation**), the corresponding alkane is formed. This reaction is carried out by heating the alkene under pressure, in the presence of a metallic catalyst. The hydrogenation of ethene to form ethane is shown below.

$$C_2H_4 \; + \; H_2 \; \rightarrow \; C_2H_6$$

ethene + hydrogen → ethane

A chemical test for alkenes is that they turn orange bromine water colourless. This happens because of an addition reaction, shown in Figure 22.5.

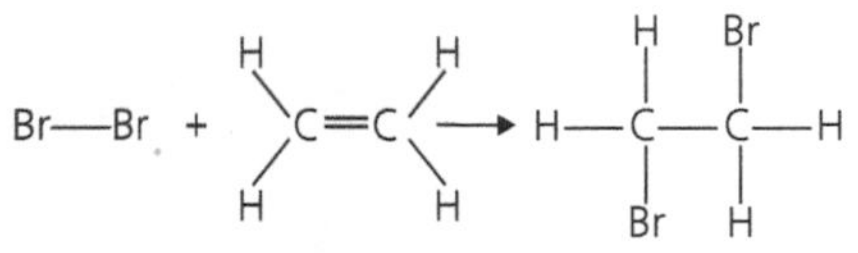

$$Br_2 \;+\; C_2H_4 \;\longrightarrow\; BrCH_2CH_2Br$$
bromine   ethene   dibromoethane

Figure 22.5 Addition reaction of ethene with bromine.

Isomerism and plastics

Isomerism in alkanes and alkenes

Two or more compounds can have the same molecular formula but different structural formulae, because the atoms are arranged in a different way. These compounds are referred to as **isomers**.

Isomerism is common in alkanes and alkenes, and examples are shown in Figure 22.6.

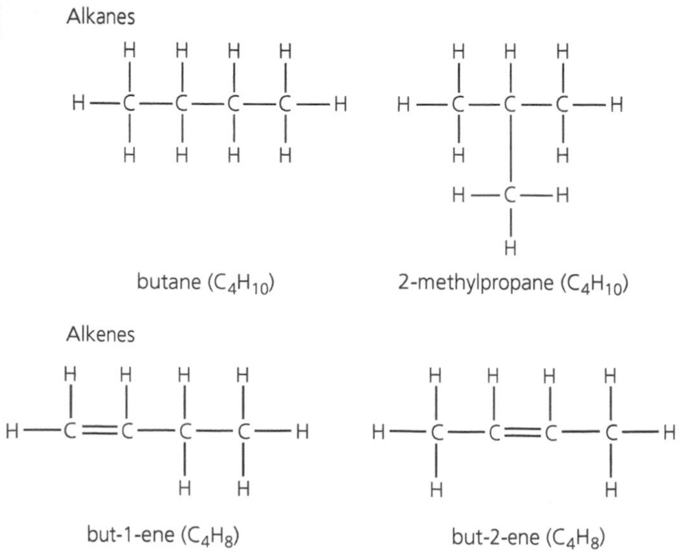

Alkanes

butane (C_4H_{10})   2-methylpropane (C_4H_{10})

Alkenes

but-1-ene (C_4H_8)   but-2-ene (C_4H_8)

Figure 22.6 Examples of isomerism in alkanes and alkenes.

In the alkane example shown, there is a difference in that in butane the carbon chain is straight, whereas in 2-methylpropane it is branched. In the alkene example the carbon-carbon double bond is in a different position. Isomers of alkanes and alkenes have the same chemical properties but slightly different physical properties (e.g. different boiling points).

Making plastics

REVISED

Alkene molecules can be assembled together in large numbers to make a variety of useful **polymers**, such as the plastics polythene, polypropene, polytetrafluoroethene (PTFE) and polyvinylchloride (PVC). A **plastic** is any synthetic or semi-synthetic organic polymer.

Plastics are one of two types.
- **Thermoplastics** are plastics which soften when heated.
- **Thermosets** are resistant to heat and do not soften or melt when heated.

Thermoplastics are used extensively for household containers such as bowls and buckets, and for packaging material. Thermosets are used for electric light fittings, saucepan handles, and other products where heat resistance is important. The difference in the behaviour of these two types of polymer can be explained in terms of their structures. Thermoplastics are made up of polymer chains that are not linked together and so can slide over one another. This results in them being easy to melt. In thermosets, the polymer chains have strong cross linkages which hold the structure together and make them resistant to heat (see Figure 22.7).

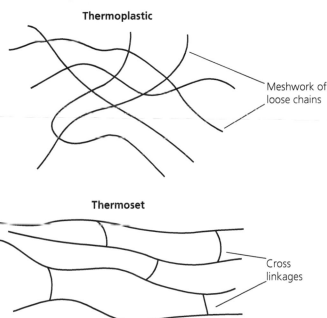

Figure 22.7 Structural differences between thermoplastics and thermosets.

Thermoset plastics are made in a liquid form. The liquid is placed in moulds and allowed to set. Once set they cannot be re-melted.

The small molecules (such as ethene) which are used to form polymers are called **monomers**. Polymers come in two groups:
- **addition polymers**, which are made from only one type of monomer
- **condensation polymers**, which are made from two or more different types of monomer.

Ethene molecules can be joined together to form the addition polymer poly(ethene), more commonly known as **polythene**. This is done by heating ethene under pressure. The number of molecules joining up varies, but is generally between 2000 and 20000 (see Figure 22.18). Polythene is commonly used for plastic carrier bags. It is flexible but not very strong. Some other addition polymers are shown in Table 22.2.

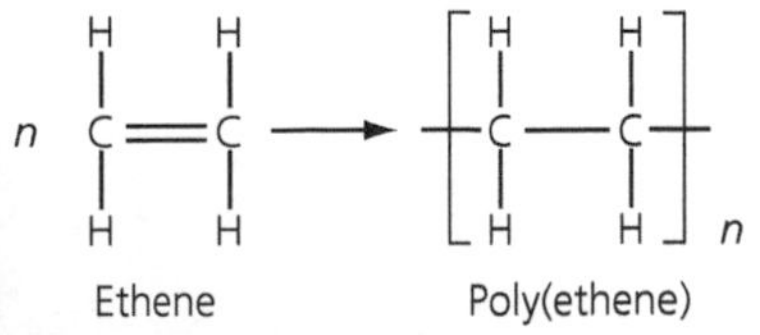

Ethene   Poly(ethene)

Figure 22.8 Formation of the addition polymer poly(ethene) from ethene (*n* is a large number).

Table 22.2 Addition polymers and their uses.

Name	Repeating unit	Uses
Polypropene	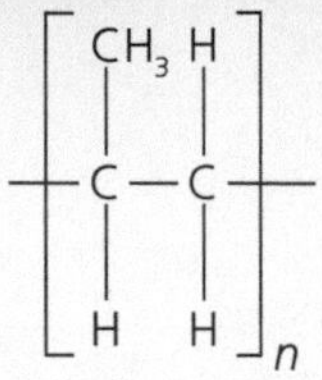	Dishwasher-safe food containers; piping; automotive parts; waterproof carpets
Polyvinylchloride PVC	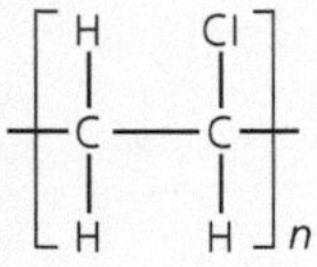	Very widely used; pipes; window and door frames; clothing; electrical insulation
Polytetrafluoroethene PTFE	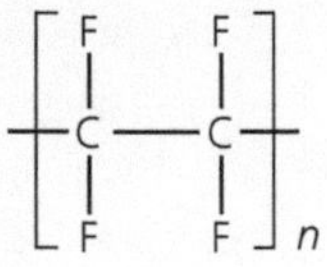	Known as Teflon; non-stick coating for cookware, irons, wiper blades, etc.

Plastics and the environment

REVISED

Many plastics are resistant to chemical attack, and **non-biodegradable**, in other words, they do not rot. Because they do not rot away, plastics tend to fill up landfill sites so that new sites have to be found. It is far better if waste plastic is recycled. This has several advantages:

- Less used plastic goes to landfill. This allows current landfill sites to be used for longer.
- Less oil is used for plastic production. Crude oil is a finite resource and needs to be conserved wherever possible.
- Less energy is consumed. This saves on costs and also reduces greenhouse gas emissions

Although it is possible to recycle all plastics, some are more difficult and expensive to recycle than others.

Now test yourself

TESTED

11 What is the general formula for
  a) an alkane and
  b) an alkene?
12 What is the test for alkenes?
13 What type of reaction converts an alkene into an alkane?
14 What is an isomer?
15 What is the difference between a thermoset plastic and a thermoplastic?

Answers online

Answers and quick quizzes at www.hoddereducation.co.uk/myrevisionnotesdownloads

Summary

- Crude oil is a complex mixture of hydrocarbons that was formed over millions of years from the remains of simple marine organisms.
- Fractional distillation of crude oil separates out fractions which can be used in a variety of ways.
- The fractions contain mixtures of hydrocarbons (alkanes) with similar boiling points.
- The compounds in the fractions have decreasing chain lengths and lower boiling points as you go up the fractionating column.
- The fractions with low boiling points and low viscosity are the most useful as fuels.
- The oil industry has global economic and political importance and social and environmental impacts.
- Hydrocarbons and other fuels undergo combustion reactions with oxygen.
- The combustion reaction of hydrogen produces no carbon dioxide.
- Hydrogen has advantages and disadvantages as a fuel.
- The fire triangle indicates the components required for fire and is used in fire-fighting and fire prevention.
- The cracking of some fractions produces smaller and more useful hydrocarbon

molecules, including monomers (alkenes) which can be used to make plastics.
- The general formula of alkanes is C_nH_{2n+2} and for alkenes is C_nH_{2n}.
- **H** Isomerism occurs in more complex alkanes and alkenes.
- There is an established system for the naming of alkanes, alkenes and other organic compounds.
- Alkenes undergo addition reactions with hydrogen and bromine.
- Bromine water is used in testing for alkenes.
- The addition polymerisation of ethene and other related monomers produces polythene, poly(propene), poly(vinylchloride) and poly(tetrafluoroethene).
- The general properties of the plastics polythene, poly(propene), poly(vinylchloride) and poly(tetrafluoroethene) give rise to different uses.
- There are environmental issues relating to the disposal of plastics, including their non-biodegradability, increasing pressure on landfill for waste.
- Recycling addresses these issues as well as the need to carefully manage the use of finite natural resources such as crude oil.

Exam practice

1 The figure shows a fractionating column used to separate crude oil into useful fractions.

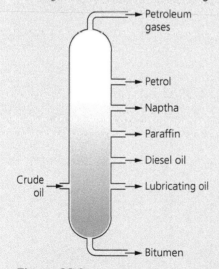

Figure 22.9

a) Which fraction would contain chemicals with the highest boiling points? [1]

b) Would the naptha fraction contain chemicals with longer or shorter chains than the diesel oil fraction? Explain the reasons for your answer. [4]

c) Apart from its boiling point, in what way would you expect the petrol fraction to differ from the diesel oil fraction, in terms of its properties? [1]

d) The longer chain fractions are broken up into smaller molecules. What name is given to this process? [1]

Answers and quick quiz 22 online

ONLINE

23 Electric circuits

This topic explores the relationship between current, potential difference (or voltage) and resistance. It shows how voltages and currents are related in series and parallel circuits, and how to calculate the total resistance a circuit. It looks at the concept of electrical **power** as the energy transferred per unit time and introduces the equations for the calculation of the power transferred by an appliance.

> **Power** is the energy transferred per unit time.

Common circuit symbols

Cell	—⊣⊢—	Ammeter	—(A)—	
Battery	—⊣⊢⋯⊣⊢—	Voltmeter	—(V)—	
Indicator lamp	—⊗—	Microphone	⊐○	
Filament lamp	—⊖—	Bell	⌓	
Switch	—o⟋o—	Buzzer	⊲	
Resistor	—▭—	Loudspeaker	⊲◁	
Variable resistor	—▱—	Motor	—(M)—	
Diode	—▷|—	LED	▷|↗↗	
Fuse	—▭—	LDR	↘↘▭	
Thermistor	—▱—	Solar cell	↘↘⊕	

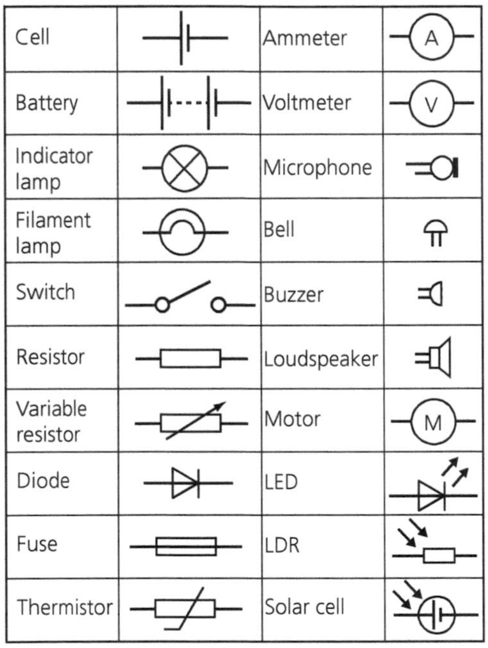

Figure 23.1 Common circuit symbols.

Current

The current flowing through electrical components in a circuit is measured in amperes (or amps), A, using an ammeter connected in series with the components.

Series circuits

For components connected in series, the current is the same at any point in the circuit. This means that all the components in a series circuit have the same current flowing through them. In Figure 23.2 the ammeter at A will read the same as an ammeter connected into the circuit at B or C.

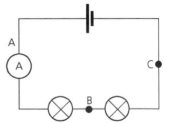

Figure 23.2 A series circuit.

Parallel circuits

When components are connected in parallel, the current splits when it gets to a junction in the circuit. No current is lost at a junction, so the total current into the junction equals the total current out of the junction. In Figure 23.3, the current at P is equal to the current at Q plus the current at R; the current at X plus the current at Y is equal to the current at Z.

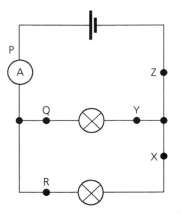

Figure 23.3 A parallel circuit.

> ### Exam tip
>
> Questions involving circuit diagrams require you to 'read' the diagram before attempting the question. Identify all the components and decide if the components are arranged in series or parallel. Pay particular attention to any labels next to components – they will have been put there for a reason!

Voltage

The voltage across components in a circuit is measured in volts using a voltmeter. Voltmeters are always connected in parallel across components, as in Figure 23.4. In series circuits, the voltages add up to the supply voltage. In parallel circuits, like Figure 23.3, the voltage is the same across each of the bulbs. A voltmeter connected between Q and Y will read the same voltage as one connected between R and X.

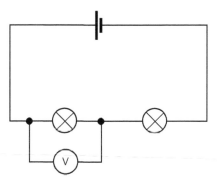

Figure 23.4 The voltmeter measures the voltage across one bulb.

Using electrical circuits in the home

Most of the mains electrical circuits in your house are connected in parallel. This has several advantages:
● If one component in the circuit stops working, all the others will continue to work properly.
● The voltage is the same for all the components.
● It is much easier to connect up all the circuits, and to add new circuits.
● It is easy to work out the total current being drawn by the different parts of the circuit (it all adds up).
● It is safer – each part of the circuit can be protected by its own fuse or circuit breaker and controlled by its own switch.

Now test yourself

1 How should an ammeter be connected into a circuit in order to measure a current?
2 A power supply provides 1.3 A to a light bulb and 0.8 A to a small electric motor connected in parallel with the light bulb. What is the total current drawn from the power supply?
3 What would be the advantage of connecting 12 Christmas tree lights in parallel, rather than in series?

Answers online

Investigating current and voltage

Figure 23.5 shows a variable resistor connected in series with a fixed resistor. The resistance of the variable resistor can be changed, in order to vary the current through, and the voltage across, the fixed resistor. The fixed resistor could be replaced by any component, such as a filament lamp, to investigate how the current and voltage vary for the component.

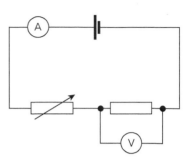

Figure 23.5 The variable resistor controls the current through, and voltage across, a fixed resistor.

Voltage–current relationships

The graphs in Figure 23.6 show the voltage–current relationships for a fixed resistor and a filament lamp.

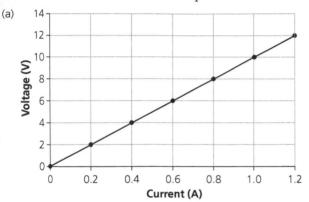

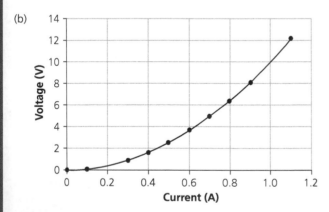

Figure 23.6 (a) Voltage against current for a 10 Ω fixed resistor and **(b)** a filament lamp.

- For fixed resistors (and wires at constant temperature), voltage and current are proportional to each other – doubling the current will double the voltage. The graph is linear (a straight line). A bigger resistance will give a bigger slope, on a V–I graph.
- For components such as filament lamps, the resistance changes with current. The resistance of a filament lamp increases with current, so the slope of the voltage–current graph increases.
- In the examination you could be shown V–I graphs (as in Figure 23.6) or I–V graphs, where current is plotted on the y-axis and V on the x-axis.

The current, voltage and resistance of electrical and electronic components are related to each other. The physicist Georg Ohm investigated this in 1827. We summarise his findings using the equation:

$$\text{current, } I \text{ (amps)} = \frac{\text{voltage, } V \text{ (volts)}}{\text{resistance, } R \text{ (ohms)}}$$

$$I = \frac{V}{R}$$

This equation can be used to calculate any one of the three variables, provided that we know the other two.

Examples

1 A 20 Ω (ohm) fixed resistor has a voltage of 12 V across it. Calculate the current through it.
2 (Higher Tier) Calculate the resistance of a filament lamp operating at 6 V with a current of 0.3 A through it.

Answers

1 $I = \dfrac{V}{R} = \dfrac{12}{20} = 0.6\,A$

2 $I = \dfrac{V}{R}$ so $R = \dfrac{V}{I} = \dfrac{6}{0.3} = 20\,\Omega$

Thermistors, diodes and light-dependent resistors

Thermistors are components rather like resistors, but their resistance changes with temperature. Most thermistors decrease their resistance with temperature – these are called negative temperature coefficient (ntc) thermistors. Thermistors can be used in circuits as electrical temperature sensors.

Light-dependent resistors (LDRs) are components that change their resistance depending on the light intensity shining on them. They can be used as light sensors in electrical circuits. Most LDRs decrease their resistance with increasing light intensity.

Diodes are electrical components that control the direction of flow of the current in a circuit. They behave like one-way electrical gates, only allowing the current to flow in one direction through the diode. Figure 23.8 shows how the electrical characteristic graph for a diode could be determined, together with an example of an electrical characteristic.

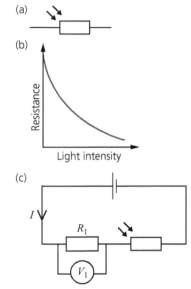

Figure 23.7 (a) The electrical symbol for an LDR. **(b)** The variation of the resistance of a typical LDR with light intensity. **(c)** An electrical circuit diagram showing how an LDR can be used in an electrical circuit.

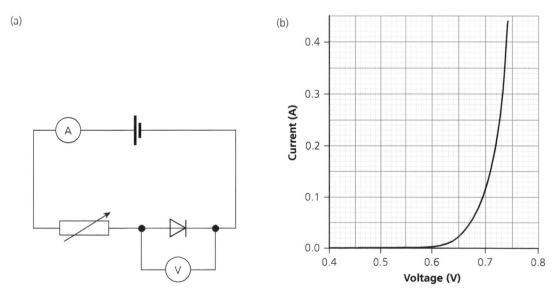

Figure 23.8 (a) A circuit for determining the I–V graph for a diode and **(b)** the electrical characteristic(I–V graph) for a diode.

Combining resistors in series and parallel

When two or more resistors (or components) are combined together in a series circuit (as shown Figure 23.9), the total resistance of the circuit increases and it is calculated from the sum of all the resistances using the equation:

$$R = R_1 + R_2$$

Combining resistors in parallel reduces the overall resistance of the circuit. Figure 23.10 shows two resistors, R_1 and R_2, arranged in parallel with a battery.

The overall resistance, R, of a parallel circuit can be calculated using the equation:

$$\frac{1}{R} = \frac{1}{R_1} + \frac{1}{R_2} + \frac{1}{R_3} + \dots$$

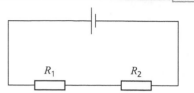

Figure 23.9 Resistors in series.

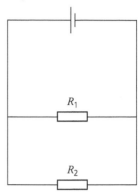

Figure 23.10 Resistors in parallel.

Electrical power

The rate of transfer of electrical energy by a device is the electrical power, P, measured in watts, W. It can be calculated using the equation:

$$P = VI$$

or the equation:

$$P = I^2R$$

Examples

1 Calculate the power of a filament lamp operating at a voltage of 12 V and a current of 0.5 A.
2 A fixed resistor with a resistance of 25 Ω has a current of 0.8 A through it. Calculate the power of the fixed resistor.

Answers

1 $P = VI = 12 \times 0.5 = 6\,\text{W}$
2 $P = I^2R = 0.8^2 \times 25 = 16\,\text{W}$

Exam tip

Often an equation is given to you in an exam question, but sometimes you are asked to find a suitable equation from the formula sheet, which will be on the inside cover of the exam paper. On the Higher Tier paper, you may also need to rearrange an equation, so you need to practise this skill.

Now test yourself

4 The V–I graph for a filament bulb is shown in Figure 23.6 (b). Sketch the I–V graph for this bulb, with I on the y-axis and V on the x-axis.
5 Calculate the current flowing through a 15 Ω resistor with a voltage of 3.0 V across it.
6 Calculate the power of the resistor in Question 5.

Answers online

Summary

- Common electrical circuit symbols are shown in Figure 23.1.
- In a series circuit, the current is the same throughout the circuit and the voltages add up to the supply voltage.

- In parallel circuits, the voltage is the same across each branch of the circuit and the sum of the currents in each branch is equal to the current in the supply.

- Voltmeters and ammeters are used to measure the voltage across, and current through, electrical components in electrical circuits.
- The voltage and current characteristics of a component can be shown on a $V–I$ graph or an $I–V$ graph.
- The equation for current I flowing through a component, where V is the voltage across a component of resistance, R, is
$$I = \frac{V}{R}$$
- Adding components in series increases total resistance in a circuit; adding components in parallel decreases total resistance in a circuit.

- The total resistance, R, of two resistors, R_1 and R_2, connected in series is given by:
$$R = R_1 + R_2$$
 ● The total resistance, R, of two resistors, R_1 and R_2, connected in parallel is given by:
$$\frac{1}{R} = \frac{1}{R_1} + \frac{1}{R_2}$$
- Electrical power is the electrical energy transferred per unit time:
$$E = Pt$$
- Electrical power can be calculated using these two equations:
$$P = VI$$
H ▶ $P = I^2R$

Exam practice

1 Figure 23.11 shows part of a mains lighting circuit that is protected by a fuse in the household fuse box (consumer unit). A, B and C are lamps; S_1, S_2 and S_3 are switches.

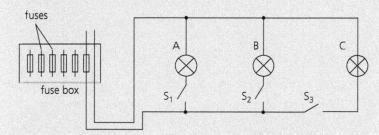

Figure 23.11

a) Copy and complete the sentences below by selecting the correct words from the choice bracket. [2]
If too much current is drawn by the lighting circuit, the fuse will melt. This makes the circuit [complete / incomplete] and the lamps will be [on / off].
b) The fuse in this circuit is working properly. For a lamp to light there must be a complete circuit.
   i) State which lamp(s) are lit when S_1 and S_2 are closed (on) and S_3 open (off). [1]
   ii) State which lamp(s) are lit when S_3 is closed (on) and S_1 and S_2 open (off). [1]
   (WJEC GCSE Physics P2 Foundation Tier Summer 2010 Q9)

2 Figure 23.12 shows part of a mains lighting circuit that is protected by a fuse in the mains fuse box (consumer unit). A, B, C and D are lamps in the circuit. The table gives information about each lamp.

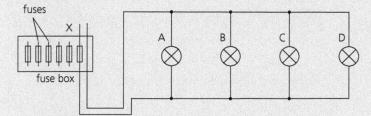

Lamp	Power (W)	Current (A)
A	40	0.17
B	60	0.26
C	40	0.17
D	60	0.26

Figure 23.12

a) When working normally, calculate how much current is flowing through the fuse at X. [1]
b) Add the following to the circuit diagram:
   i) a switch labelled S_1 which controls lamp A only
   ii) a switch labelled S_2 which controls lamps C and D only. [2]
   (WJEC GCSE Physics P2 Higher Tier Summer 2010 Q4)

3 Figure 23.13 shows an ammeter A and a voltmeter V connected to a power supply and a resistance wire XY. A connector S allows the length of wire in the circuit to be changed.

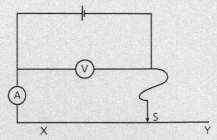

Figure 23.13

a) With S in the position shown, the voltmeter reads 6 V and the ammeter 1.2 A. State a suitable equation that could be used to calculate the resistance of the wire between X and S, and then use the equation and the data to calculate this. [3]

b) The connector S is moved towards Y. State the effect, if any, this would have on:
   i) the resistance in the circuit [1]
   ii) the ammeter reading. [1]

(WJEC GCSE Physics P2 Higher Tier Summer 2010 Q1)

4 The circuit shown in Figure 23.14 is used to investigate how the resistance of a lamp changes.

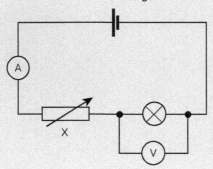

Figure 23.14

a) Explain how component X allows a set of results to be obtained. [2]

b) The results obtained are used to plot the graph shown in Figure 23.15.

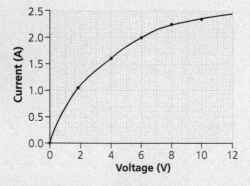

Figure 23.15

i) Write down in words an equation from the equations list on page viii and use it to calculate the resistance of the lamp when the voltage across it is 4 V. [4]

ii) Use the graph and a suitable equation from the equations list on page 253 to calculate the power of the lamp when the voltage across it is 4 V. [3]

(WJEC GCSE Physics P2 Higher Tier Summer 2008 Q6)

Answers and quick quiz 23 online

ONLINE

24 Generating electricity

Electricity is a very useful form of energy, as it can be produced in large quantities. Also, it is easily and efficiently transferred around the country using transformers and the National Grid. Electricity can be transferred into other useful forms, such as light and heat, with relative ease. Electricity can be generated using **renewable** or **non-renewable technologies**, each with its own set of advantages and disadvantages. Over 90 per cent of the UK's electricity is generated in large power stations using fossil fuels, such as coal, oil or gas, or by using nuclear power.

Renewable energy technologies use resources that will never run out, because these can be replenished.

Non-renewable technologies use energy resources that will run out, because there are finite reserves which cannot be replenished.

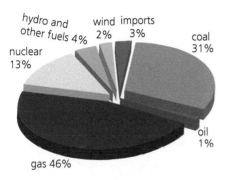

Figure 24.1 Generation of electricity in the UK by fuel type.

Exam tip

Always study graphs, diagrams and charts very carefully before answering questions based on them.

Advantages and disadvantages of generating electricity from different primary sources of energy

REVISED

Primary source of energy	Advantages	Disadvantages
Fossil fuels (such as coal, oil and gas)	Large amounts of electricity can be generated cheaply. Fossil-fuel power stations are very reliable. There is security of the supply of fossil fuels.	Fossil-fuel power stations can be dirty (especially coal). Burning fossil fuels produces carbon dioxide gas which contributes to the greenhouse effect and global warming. Burning fossil fuels produces sulfur dioxide gas, which contributes to acid rain. Large amounts of fuel need to be brought onto site and waste (in the case of coal) needs to be removed from the site. Fossil fuels are non-renewable forms of energy.

→

Primary source of energy	Advantages	Disadvantages
Nuclear energy	Does not give out **greenhouse gases** (no air pollution). It can produce energy for long periods of time without the need for refuelling. It is very reliable. It can produce lots of energy.	It can be expensive to build a nuclear power station. It is expensive to decommission a nuclear power station. Radioactive waste needs to be stored securely for a very long time. Nuclear power is non-renewable. There is risk of terrorist attack. There is a danger of potential nuclear accident.
Wind energy	Wind does not use up a fuel. It is a renewable form of energy. There is no air pollution.	Windy sites tend to be away from centres of population – high-voltage power lines, which are unsightly, are needed to transmit the electricity. Wind turbines only operate when it is windy. Each wind turbine can only produce a small amount of electricity, so many turbines are required. Wind turbines can be unsightly.
Solar power	Renewable form of energy. Readily available and predictable – in daytime. Cheap to install. Solar panels can be retro-fitted to buildings. Easy to install in areas with large populations.	Does not generate electricity at night. Large-scale solar power stations use up a lot of land. Large areas of solar panels are needed to generate large amounts of electricity.
Hydroelectric power (HEP)	HEP is renewable. There is no air pollution. Large HEP stations can generate enormous amounts of electricity reliably. HEP stations have almost instant start-up times, so can be switched on and off easily. There are no fuel costs. No fossil fuels are used.	Large dams need to be constructed, which can be expensive to build. Valleys are flooded when dams are constructed, destroying habitats. Suitable HEP sites tend to be away from centres of population – high-voltage power lines, which are unsightly, are needed to transmit the electricity. Drought can reduce the supply of water needed to produce HEP.
Wave and tidal energy	Both wave and tidal energy are renewable energies. Tidal energy is very predictable. Large-scale tidal power stations could generate huge amounts of electricity. No fossil fuels are used. Non-polluting. Both forms of power generation could be turned on and off very quickly.	Wave energy is unreliable and only works when there are suitable waves. Large numbers of wave generators would be needed to generate significant amounts of energy. Tidal energy barrages cause large-scale flooding of estuaries, destroying habitats.

➜

Primary source of energy	Advantages	Disadvantages
Biofuels (such as animal waste, wood and fast-growing crops)	Biofuels are a renewable form of energy. Large-scale biofuel power stations could be built, generating large amounts of electricity.	Large areas of land are needed for fast-growing plants/trees, or a large amount of animal waste is needed, which would have to be transported cleanly. Although carbon neutral, carbon dioxide is still emitted into the atmosphere. Biofuel power stations can be unsightly.
Geothermal energy	Renewable form of energy. Does not produce any pollution. Reliable source of energy in places where there are hot springs or where hot rocks are close to the surface. Ground-source heat pumps can be installed into domestic houses. Cheap form of energy.	Hot springs and hot rocks are only available in certain areas, normally away from large populations, so unsightly pylons and cables are needed. Ground-source heat pumps need a lot of area to capture heat.

Exam tip

Extended writing questions require you to produce good quality written answers; you must be careful about the quality of the written communication that you use. Obvious things to check are: logical organisation of your thoughts and arguments; punctuation and grammar, such as correct use of capital letters and full stops; and correct use and spelling of key scientific terms.

A **greenhouse gas** is one that traps radiation in the Earth's atmosphere and, therefore, contributes to global warming. Carbon dioxide and methane are examples of greenhouse gases.

Comparing the cost

REVISED

Cost	Coal power station	Wind farm	Nuclear power station
Commissioning costs: Buying land Professional fees Building costs Labour costs	High	Low	Very high
Running costs: Labour costs Fuel costs	High	Very low	High
Decommissioning costs: Removal of fuel (nuclear) Demolition Clean-up	High	Low	Very high

1 Compare the proportion of UK electricity generated by gas to the proportion generated by wind power.
2 What advantages does generation of electricity from fossil fuels have over nuclear power?
3 State some forms of renewable energy that are unreliable.
4 Explain why generating electricity using nuclear power is still used, despite the huge costs involved.

Answers online

Generating electricity in a fuel-based power station

REVISED

In a (fossil) fuel-based power station, the chemical energy stored within the coal, oil or gas is burnt in a furnace at several thousand degrees Celsius, together with air/oxygen, producing enough thermal heat energy to turn large amounts of water into steam every second in the boiler. The steam is then superheated and pressurised, causing it to have a huge amount of kinetic energy. This then turns turbines, spinning them at several thousand revolutions per minute (rpm). Each of the turbines is connected to an electric generator, producing large amounts of electrical energy, which is output to the National Grid. The superheated steam is then cooled and condensed back into water, by passing it through kilometres of pipes inside cooling towers.

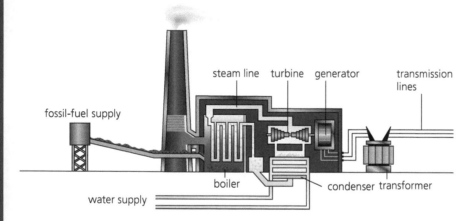

Figure 24.2 A schematic diagram of a typical fossil-fuel power station.

In a nuclear power station, a nuclear reactor is used to produce the heat needed to turn the water into superheated steam.

Why do we need the National Grid?

REVISED

Over 90 per cent of the UK's electricity is generated in large-scale power stations. The amount of electricity produced by these power stations is controlled by the National Grid, which provides:
● a reliable, secure energy supply
● an electricity supply that matches the changing demand during the day and over the course of the year
● high-voltage power lines that connect power stations to consumers
● electrical substations that control the voltage being supplied to consumers.

The amount of electricity consumed over the course of one day and over the course of the year varies in very predictable ways:

- Peak daily consumption is around 6 p.m., when people are cooking evening meals.
- Overall consumption is higher in the winter than in the summer, as people use more electricity for lighting and heating.

The National Grid

When electrical current passes down a wire, it causes the wire to heat up. The heat energy generated from the electricity then transfers into the surroundings, heating up the air. The larger the current, the greater the heat loss.

The National Grid is designed to minimise the amount of energy lost as heat when electricity passes down the power lines. The electricity generated at power stations is changed by step-up transformers to very high voltages (typically 400000V, 275000V or 132000V) but very low current – so that the energy lost as heat in the power lines is very small. (Only about 1 per cent of the total energy transmitted is lost in this way.)

High voltages would be very dangerous if used in homes and offices, so step-down transformers change the electricity to a lower voltage and higher current for use by consumers.

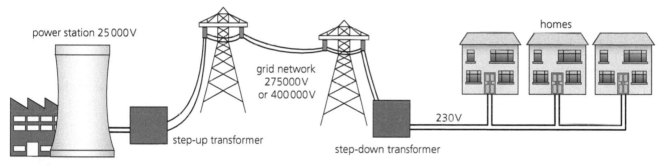

Figure 24.3 The National Grid transmission system.

Electrical power

REVISED

Electrical power is a measure of the rate at which electrical energy can be transformed into other more useful forms of energy. Electrical power is calculated using the equation:

electrical power = voltage × current

$$P = VI$$

In most UK houses, mains voltage $V = 230\,V$.

Example

A mains hairdryer draws a current of 5.5 A. Calculate the power of the hairdryer.

Answer

Mains voltage = 230 V

Hairdryer current = 5.5 A

$$P = VI$$

power = 230 × 5.5 = 1265 W

Exam tip

When you are asked to do calculations involving units with prefixes (such as kV or MW) make sure that you convert the numbers carefully back into base numbers. For example, 400 kV = 400000V and 100 MW = 100000000W.

TESTED

Now test yourself

5 State the useful energy transfers within a fossil-fuelled power station.
6 Give two reasons why we need a National Grid.
7 Why is electricity transferred around the National Grid at high voltage and low current?
8 Calculate the power of a mains kettle operating at a voltage of 230 V and a current of 13 A.
9 Calculate the current flowing through a 2.5 kW mains lawnmower operating at a voltage of 230 V.

Answers online

> **Exam tip**
>
> *Calculate* means that you must produce a numerical answer by doing a mathematical calculation.

Sankey diagrams

REVISED

The transfer of energy (or power) from one form to other forms can be shown using a Sankey diagram, which shows the types and amounts of energy as they transform into different forms. The Sankey diagram for an energy-efficient light bulb is shown in Figure 24.4.

Sankey diagrams are drawn to scale – the width of the arrow at any point shows the amount of energy being transformed. Conventionally, we write the type of energy and the amount of energy (or power) on the arrow, and the useful forms of energy usually go along the top of the diagram, with the wasted forms curving off downwards. Sankey diagrams give us not only a good way of showing energy (and power) transfers by a device or during a process, but also an indication of how efficient the process is – the bigger the useful energy arrow is compared to the input arrow, the higher the efficiency.

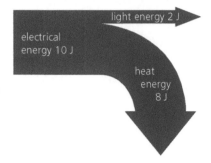

Figure 24.4 Sankey diagram for an energy-efficient light bulb.

Efficiency

REVISED

Efficiency is a measure of how much useful energy (or power) comes out of a device or process compared to the total amount of energy (or power) that goes into the device or process. Efficiency is usually expressed as a percentage using the equation:

$$\% \text{ efficiency} = \frac{\text{useful energy (or power) transfer}}{\text{total energy (or power) input}} \times 100$$

Example

The efficiency of an energy-efficient light bulb can be calculated using data from the Sankey diagram in Figure 24.4. The total energy input (as electricity) is 10 J. The useful energy output (as light) is 2 J.

Answer

$$\% \text{ efficiency} = \frac{\text{useful energy transfer}}{\text{total energy input}} \times 100$$

$$= \frac{2}{10} \times 100 = 20\%$$

> **Exam tip**
>
> Foundation-tier students are not required to rearrange equations – but you might have to change numbers given with prefixed units, such as kW, back into base units, such as W.

Why is energy efficiency important?

Energy-efficient devices are very important for the future. The more efficient a device is, the more of the energy input is output as useful energy and less is wasted. Conventional fossil-fuel power stations are at best only 33 per cent efficient. This means that for every 100 tonnes of coal or oil used, only about 33 tonnes is converted directly into useful electricity. The rest of the coal or oil is effectively heating up the atmosphere and producing unnecessary carbon dioxide gas. Wind turbines are about 50 per cent efficient and solar panels about 30 per cent efficient. Standard tungsten filament light bulbs are typically only 2–3 per cent efficient; 'low-energy' bulbs are about 20 per cent efficient but LED light bulbs can be up to 90 per cent efficient. Imagine the effect on electricity consumption if every light bulb in the UK was replaced by an LED bulb.

Exam tip

You will come across two types of equation-based questions in the examination. You may be given an equation and asked to extract the correct data from the question to use in the equation; or you will be given the correct data and you will be asked to select the appropriate equation from the list at the front of the examination paper.

On the Higher Tier paper, you may also need to rearrange the equation as part of your answer.

Now test yourself

TESTED

10 Draw the Sankey diagram for 400 MW output gas-powered station, where 1000 MW of chemical energy is input from the gas.
11 Calculate the efficiency of the gas-powered power station in Question 10.
12 An 18 W LED light bulb is 90 per cent efficient. Calculate the power output as light.

Answers online

Summary

- Electricity is a really useful form of energy because it is easy to produce and easy to transfer into other useful forms of energy.
- Power stations (renewable, non-renewable and nuclear) have significant but very different commissioning costs, running costs (including fuel) and decommissioning costs that need to be considered when planning a national energy strategy.
- Large-scale power generation by power stations and micro-generation using renewable technologies, e.g. using domestic wind turbines and roof-top photovoltaic cells, have different advantages and disadvantages. They all have very different environmental impacts.
- Data can be used to determine the efficiency and power output of power stations and renewable energies.
- In a fuel-based power station, electricity is generated by the burning fuel, producing thermal energy which boils water to produce steam. The moving steam turns a turbine, attached to a generator, which produces the electrical energy.

- Sankey diagrams can be used to show energy transfers.
- The efficiency of an energy transfer can be calculated from the equation:

$$\% \text{ efficiency} = \frac{\text{useful energy (or power) transfer}}{\text{total energy (or power) input}} \times 100$$

- There is a need for a national electricity distribution system (the National Grid), in order to maintain a reliable energy supply that is capable of responding to a fluctuating demand.
- The National Grid consists of power stations, substations and power lines.
- Electricity is transmitted across the country at high voltage because it is more efficient, but low voltage is used at home because it is safer.
- Transformers are needed to change the voltage and current within the National Grid.
- It is possible to experimentally investigate the operation of step-up and step-down transformers, in terms of the input and output voltage, current and power.
- power = voltage × current

  $P = VI$

Exam practice

1 Study Figure 24.1 showing the proportions of electricity generated by different fuels. Calculate the percentage of UK electricity generated from fossil fuels and nuclear power combined. [1]

2 Discuss the factors that are involved in making decisions about the type of commercial power station that could be built in an area. [3]

(WJEC GCSE Physics P1 Higher Tier Summer 2010 Q5(a))

3 Large coal-fired power stations are generally built close to lakes or rivers and near to both motorways and mainline railways. Suggest why coal power stations:
   a) require good road and railway links [1]
   b) are built near a source of water. [1]

(WJEC GCSE Physics P1 Higher Tier January 2009 Q4(b))

4 If you live on the coast of Britain, the area may be ideal for building a power station nearby. The choice may be between building a nuclear or a coal-fired power station.
   a) People often object to power stations because of their appearance. Write a paragraph describing three other objections you could raise to nuclear power stations. [3]
   b) Write a paragraph describing three objections you could raise, apart from appearance, to coal-fired power stations. [3]

(WJEC GCSE Physics P1 Higher Tier Summer 2008 Q6)

5 The table shows some of the information that planners use to help them decide on the type of power station they will allow to be built.

	Wind	Nuclear
Overall cost of generating electricity (p/kWh)	5.4	2.8
Maximum power output (MW)	3.5	3600
Lifetime (years)	15	50
Waste produced	None	Radioactive substances, some of which remain dangerous for thousands of years
Lifetime carbon footprint (g of CO_2/kWh)	4.64 (onshore) 5.25 (offshore)	5

   a) Give one reason why the information in the table does not support the idea that wind power will be a cheaper method of producing electricity. [1]
   b) Supporters of wind power argue that it will reduce global warming more than nuclear power. Explain whether this is supported by information in the table. [2]
   c) Supporters of nuclear power argue that it will meet a greater demand for electricity in the future than wind power. Give two ways in which this is supported by information in the table. [2]

(WJEC GCSE Physics P1 Higher Tier January 2010 Q2)

6 Figure 24.5 shows part of the National Grid. Electricity is generated at power station A.

Figure 24.5

   a) Use the words below to copy and complete the sentences that follow. Each word may be used once, more than once or not at all.

   transformer   pylon   generator   power   current

   i)   At B, a _____ increases the voltage. [1]
   ii) Electricity is sent at a high voltage along C, so the _____ is smaller. [1]
   iii) At D, the voltage is decreased using a _____. [1]
   b) Explain why the electricity is stepped up at B, but stepped down at D. [3]

→

c) Assume that electricity is transmitted along the cables C at a power of 100 MW and a voltage of 400 kV. Use the equation: power = voltage × current to calculate the current in the cables. [3]

(WJEC GCSE Physics P1 Foundation Tier January 2010 Q7)

> **Exam tip**
>
> Read the instructions in questions carefully. Note that, in Question 6 (a) you can use any word from the list more than once and some words may not be used at all.

7 Figure 24.6 shows part of the National Grid.

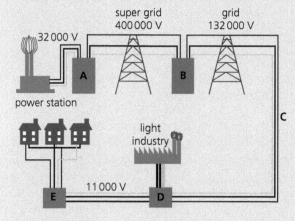

Figure 24.6

a) At which point, A, B, C, D or E, would you find a step-up transformer? [1]
b) What is the voltage at point C? [1]
c) Where is the voltage stepped down to 230 V; give the letter A, B, C, D, or E? [1]
d) Select the correct letter. A high voltage is used in the National Grid so that the electrical energy lost in the cables is:
   A   zero
   B   small
   C   big. [1]

(WJEC GCSE Physics P1 Foundation Tier January 2011 Q5)

8 Water can be boiled using a saucepan on a gas-cooker ring. The energy transfers are shown in Figure 24.7.

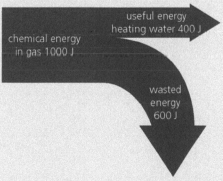

Figure 24.7

a) Write down an equation and use it to find the efficiency of heating water in this way. [3]
b) An electric kettle is 90 per cent efficient at boiling water. Copy and complete the energy transfer diagram in Figure 24.8. The diagram is not to scale. [2]

→

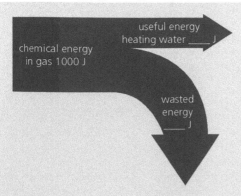

Figure 24.8

(*WJEC GCSE Physics P1 Higher Tier Summer 2010 Q3*)

9 The table shows how energy is used in a coal-burning power station. Write down in words a suitable equation and use it to calculate the efficiency of the power station.   [3]

Energy input per second	Energy output per second
6000 MJ	3350 MJ of energy is taken away as heat in the water used for cooling
	2100 MJ of energy is fed into the National Grid
	550 MJ of energy is given out in the gases released during burning

(*WJEC GCSE Physics P1 Higher Tier January 2009 Q4(a)*)

Answers and quick quiz 24 online

ONLINE

25 Making use of energy

Conduction, convection and radiation REVISED

Homes are heated by transforming energy sources such as electricity or gas into heat using appliances such as electric fires or hot-water radiators. Thermal (heat) energy will move from somewhere hot (where the temperature is higher) to somewhere cold (where the temperature is lower). It does this by conduction, convection or radiation.

Conduction

Conduction is the transfer of energy from hot to cold by the successive vibration of particles within solids and liquids. Materials that do not conduct thermal energy very well are called insulators – many non-metals are good insulators.

H▶ Materials such as metals are very good thermal conductors because they have free mobile electrons within their structure.

Convection

Convection occurs through liquids and gases. When a gas (or liquid) is heated, the particles move faster. As the particles speed up, they get further apart, increasing the volume of the gas. This causes the density of the gas to decrease. Less dense gas floats (or rises) above denser gas. As the gas rises it cools again, the particles slow down, get closer together and fall, increasing the density of the material. This creates a convection current which can heat a room. Temperature differences within the Earth's mantle and within the atmosphere cause natural convection currents.

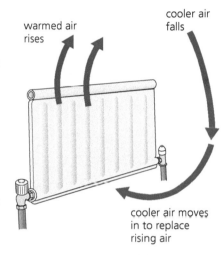

warmed air rises

cooler air falls

cooler air moves in to replace rising air

Figure 25.1 Convection currents transfer heat from the radiator to the room.

Radiation

Thermal radiation is emitted by hot objects. A hot-water radiator emits infrared electromagnetic radiation. Dull, black objects are good emitters and absorbers of thermal radiation. Shiny, light-coloured objects are good reflectors of thermal radiation. All objects emit thermal radiation, but the higher the temperature of the object the greater the amount of thermal radiation emitted.

Applying density REVISED

The density of a material can be calculated using the equation:

$$\text{density} = \frac{\text{mass}}{\text{volume}}$$

Solids have high densities because the particles inside them are generally closely packed together in a regular shape. Liquids have relatively high densities (but less than the corresponding solid) because the particles are still close packed (but free to move over each other). Gases have low densities as the particles are far apart.

Density is important when it comes to generating energy from several renewable sources such as wind energy.

Example

Using wind energy

If:

- area of turbine blades is 25 m²
- peak wind speed is 12 m/s
- density (of air) = $\dfrac{\text{mass (of air)}}{\text{volume (of air)}}$ = 1.2 kg/m³

Then:

volume of air moving through turbine blades per second =

speed × area = 12 m/s × 25 m² = 300 m³

mass of air moving through turbine per second =

density × volume = 1.2 kg/m³ × 300 m³ = 360 kg

kinetic energy of wind moving through turbine per second =

$KE = \dfrac{1}{2}mv^2 = 0.5 \times 360 \times 12^2 = 25\,920\,J$

And if the input power of wind is 25 920 W and the efficiency of the turbine is 25 per cent, then output power of turbine =

$25\,920\,W \times \dfrac{25}{100} = 6480\,W = 6.48\,kW$

Now test yourself

TESTED

1 Explain how thermal energy could be transferred away from a desk lamp.
2 Explain why metals are such good conductors of thermal energy.
3 Calculate the density of a 6 m³ block of concrete that has a mass of 14 400 kg.
4 Calculate the mass of water per second that moves through a water turbine if 0.5 m³ passes through per second and water has a density of 1000 kg/m³.

Answers online

Insulation

REVISED

The amount of thermal energy escaping from a house can be reduced by using domestic insulation systems that work by reducing the effects of **thermal conduction**, convection and radiation. The table summarises the main systems that can be installed.

> **Thermal conduction** is the flow of heat energy through a material. Metals have very high conductivities. Brick and glass have lower conductivities, but energy still flows through them fast enough to cool a house down on a cold day.

The payback time for heating or insulation system is given by the equation:

$\text{payback time} = \dfrac{\text{installation cost}}{\text{annual savings}}$

Insulation system	How it works	Typical installation costs	Typical annual savings	Payback time (years)
Draught proofing	Draught excluders and draught-proofing strips are fitted, reducing the convection of hot air through gaps under doors and in window frames.	£50	£50	1
Cavity-wall insulation	Fills the space between the double walls of bricks with foam. The foam traps air, which is a poor conductor and prevents air from circulating within the cavity, reducing thermal loss by convection.	£250	£110	2.3
Floor insulation	Mineral wool is laid between the joists under the floorboards and silicone sealant is used to seal gaps between skirting boards and floorboards. This reduces thermal loss via conduction and convection.	£140	£70	2
Loft insulation	Mineral wool insulation is laid between the timber joists in the loft. This reduces thermal loss via conduction and convection.	£250	£150	1.7
Double glazing	Two sheets of glass with a gap between them. Reduces thermal loss via conduction and convection.	£2000	£130	15.4

Units of energy – the kilowatt-hour

REVISED

Home energy values are usually compared in units equivalent to the **kilowatt-hour**, kWh.

1 kWh is equal to the amount of heat energy produced by a 1 kilowatt (1000 W) electric fire in one hour (3600 s).

$$1 \, \text{kWh} = 1000 \, \text{W} \times 3600 \, \text{s} = 3\,600\,000 \, \text{J}$$

> The **kilowatt-hour** is a unit of energy used in the domestic context.

Heating and transport costs

REVISED

Homes can be heated using a variety of different fuels. The table summarises some of the costs involved.

Fuel	Fuel price (p per unit)	Unit	Cost per kWh of fuel (p)	Energy content (kWh per unit)	CO_2 emissions per kWh
Electricity	16.8	kWh	16.8	1.0	0.5
Gas	4.7	kWh	5.2	1.0	0.2
Oil	54.1	litre	5.8	10.4	0.3
LPG (liquid propane gas)	36.7	litre	6.1	6.7	0.2
Butane	137.0	litre	19.1	8.0	0.2
Propane	74.2	litre	11.7	7.1	0.2
Wood	20.8	kg	5.8	4.2	0.03
Coal	30.0	kg	5.8	6.9	0.4

The costs associated with transport are quite variable, as they are very dependent on the world wholesale energy prices. A simple comparison table compares three variations of the Renault Clio/Zoe car; one diesel, one petrol and the electric Zoe.

Car (fuel)	Cost to buy	Road tax	CO_2 emissions (g/km)	Average fuel cost per year
Renault Clio Expression+ TCe ECO (petrol)	£13 245	£120	99	£1028
Renault Clio Expression+ dCi ECO (diesel)	£14 345	£100	83	£740
Renault Zoe Expression	£13 650	Zero	Zero	£1006 (including battery rental)

Now test yourself

TESTED

5 Which thermal energy transfer process is reduced by fitting draught excluders?
6 A house is heated with 54 MJ of thermal energy. Calculate the number of kWh of thermal energy used.
7 A builder is researching which fuel would be best for a new-build house. Use the table of data on page 198 to decide which fuel would give the cheapest running costs.
8 The builder in Question 7 wants to install cavity-wall insulation into the new-build house. The insulation will cost £350 to install and the insulation manufacturer has quoted a payback time of 2.5 years. Calculate the typical annual savings of this system.

Answers online

Summary

- Temperature differences lead to the transfer of energy thermally by conduction, convection and radiation.
- The density of an object is given by the equation:

$$\text{density} = \frac{\text{mass}}{\text{volume}}$$

- Differences in density occur between the three states of matter, due to the different arrangements of the atoms or molecules.
- Energy loss from houses can be restricted by systems such as: loft insulation; double glazing; cavity-wall insulation and draught excluders.
- The cost effectiveness and efficiency of different methods of reducing energy loss from the home can be compared in order to test their effectiveness. This can include calculating the payback time and the economic and environmental issues surrounding controlling energy loss.
- Data can be used to investigate the cost of using a variety of energy sources for heating and transport.

Exam practice

1 A homeowner decided to reduce their heating bill by improving their house insulation. The table below shows the cost of the improvements made and the yearly savings.

Insulation method	Cost	Yearly saving
Draught-proofing doors and windows	£80	£30
Fitting a jacket to the hot-water tank	£20	£20
Cavity-wall insulation	£1100	£50
Loft insulation	£400	(i)
Total	(ii)	£200

   a) Copy and complete the table by giving values for (i) and (ii). [2]

   b) The homeowner spent £1200 per year heating his house before insulating it. How much would he expect to spend each year after the improvements? [1]

   c) Give a reason why heat loss by convection is reduced by cavity-wall insulation. [1]

(WJEC GCSE Physics P1 Foundation Tier Summer 2010 Q4)

2 a) State how double glazing reduces the amount of heat lost through the windows of a house. [2]

   b) The graph in Figure 25.2 shows the results of an investigation to see how the rate of loss of energy through a double-glazed window was affected by the width of the air gap between the two panes of glass. The investigation used a window of area $1\,m^2$ and kept a temperature difference of $20\,°C$ between the inside and the outside.

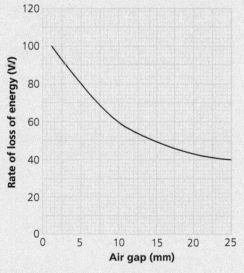

Figure 25.2

   i) Use the graph to estimate the rate of loss of energy for an air gap of 0 mm, and explain how you obtained your answer. [2]

    →

   ii) Give two reasons why most manufacturers of double-glazed windows are unlikely to use an air gap any larger than 20 mm. [2]

(WJEC GCSE Physics P1 Higher Tier Summer 2009 Q4)

3 A water turbine is sited in a river flowing at 2 m/s. The density of water is 1000 kg/m³ and 0.15 m³ of water passes through the turbine per second.

   a) Calculate the mass of water flowing through the turbine per second. [2]

   b) The water turbine produces an electrical output of 48 W. The water inputs 120 W of kinetic energy. Calculate the efficiency of the water turbine. [1]

4 A householder buys gas for heating and cooking, and electricity for lighting and operating electrical appliances. The table shows information about the householder's energy consumption and the total yearly cost.

Year	Units of electricity (kWh)	Units of gas (kWh)	Total units of energy (kWh)	Total cost (£)
1st Jan–31st Dec 2015	4309	36958	41267	866.62
1st Jan–31st Dec 2016	4540	33446	37986	949.65

   a) Use the data from the table to find the overall cost of 1 unit (kWh) of energy in 2016. [3]

   b) On 1st January 2016 the householder fitted a solar panel, at a cost of £2000, to provide hot water for heating.

      i) Use data from the table to estimate the number of units produced by the solar panel in 2016. [1]

      ii) Use the answer from part (a) to calculate the amount of money he saved on his 2016 gas bill. [1]

      iii) Calculate the time it would take for his annual savings to pay back the cost of the solar panel. [2]

      iv) Give a reason why the payback time calculated in (iii) could be much smaller. [1]

(WJEC GCSE Physics P1 Higher Tier January 2009 Q7)

Answers and quick quiz 25 online

ONLINE

26 Domestic electricity

How much does it cost to run?

The cost of running an electrical device depends on the electrical power of the device (given in kW, where 1 kW = 1000 W) and the electricity tariff (the cost per unit of electrical energy), and is calculated as follows:

energy transfer = power × time

or

$E = P \times t$

The cost of domestic electricity energy consumed is calculated using kilowatt-hours (kWh) or units. 1 kWh is equivalent to the amount of electrical energy used by a standard 1 kW electric fire in 1 hour. Units of electrical energy are calculated using the equation:

units used (kWh) = power (kW) × time (h)

The cost of the electrical energy is then given by:

cost = units used × cost per unit

The power rating of an electrical appliance is written on a plate attached to the appliance, and most domestic appliances in the UK are sold with an energy banding value (A–G), which tells you how efficient the appliance is. This is either written on the box that the appliance was sold in, or is attached as a sticker or a tag.

> **Exam tip**
>
> Be careful when calculating electrical energy consumption in kilowatt-hours. The time must be in hours. If you are given a time in minutes, you must convert it to hours first.

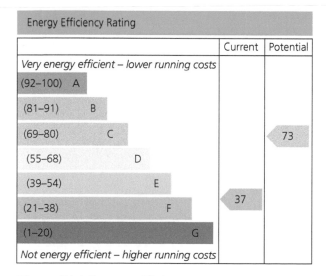

Figure 26.1 Energy efficiency rating label for an electrical appliance.

1 A 200 W lamp is left on for 15 minutes. How much electrical energy is transferred?
2 A 2 kW heater is left on in your room. You put it on at 6 a.m. and forget about it until 5 p.m. If a unit (1 kWh) costs 15p, how much will it have cost to leave the heater on?

Answers online

a.c. or d.c.?

REVISED

Electricity can either be direct current, d.c., whereby the electric current flows only in one direction, or it can be a.c., alternating current, whereby the electric current flows in one direction for half a cycle, then in the other direction for the rest of its cycle. An oscilloscope can be used to show the difference between the two types of current. Example traces are shown in Figure 26.2.

Power supplies such as batteries and solar cells produce d.c. and generators produce a.c. In the UK, a.c. current has a frequency of 50 Hz.

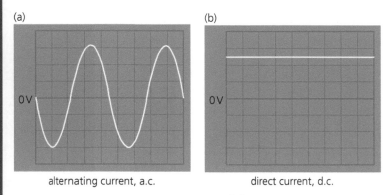

(a)   (b)

0 V   0 V

alternating current, a.c.   direct current, d.c.

Figure 26.2 Oscilloscope traces for (a) alternating current and (b) direct current.

Using mains electricity safely

REVISED

Domestic electricity is supplied to your house at 230 V with a maximum current of about 65 A. The distribution of the electric current around your home is controlled by a consumer unit, consisting of the supply connection and a series of circuit breakers that control the amount of electric current being supplied to each individual circuit around the house.

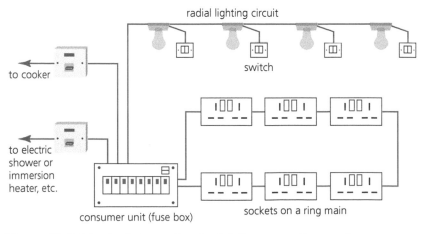

radial lighting circuit

switch

to cooker

to electric shower or immersion heater, etc.

consumer unit (fuse box)

sockets on a ring main

Figure 26.3 A typical domestic electricity connection system.

The consumer unit not only distributes the electric current around the house, it is also a safety system, preventing too much electric current from being drawn from the supply and cutting off any circuits that are faulty (short circuit). Consumer units are a series of circuit breakers, consisting of miniature circuit breakers (mcb) and residual current circuit breakers (rccb), as shown in Figure 26.4.

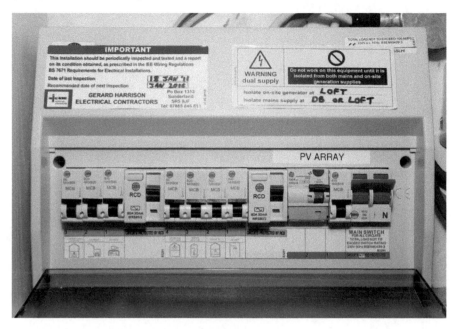

Figure 26.4 An electricity consumer unit.

Mcb switches control individual circuits within the house, such as the lighting circuits. They allow circuits to be turned on or off, and also limit the amount of current being drawn. If a short circuit occurs, the current being drawn from the consumer unit rises very quickly, exceeding the value of the mcb current rating. This turns off the circuit, isolating it from the supply and making the circuit safe. Rccb switches monitor the difference between the current drawn from the consumer unit and the current returning to it. If the difference between the two currents exceeds the rating on the rccb, it switches off, isolating the circuit from the supply. Rccb switches reduce the risk of electrocution to the people in the house because, if someone accidently touches the live wire of any part of the circuit, a larger current will be drawn from the consumer unit than returns to it, triggering the rccb and breaking the circuit. Mcb and rccb switches can be reset.

> **Exam tip**
>
> Mcb and rccb switches both isolate the electricity supply. Rccb switches work very quickly so they reduce the risk of electrocution.

Fuses

Fuses are like circuit breakers in appliances in that, if the current exceeds the rating (or maximum current) of the fuse, the wire within the fuse quickly heats up and melts, disconnecting the appliance from the supply. Fuses are fitted to the live wire of the plug and they prevent too much current from flowing through an appliance – which could cause a fire. Unlike mcb switches, standard cartridge fuses cannot be reset; they have to be replaced.

Figure 26.5 A standard 13 A fuse and its circuit symbol.

The ring main

The main socket circuit within a house is called the ring main. Only one cable is needed to connect all the sockets in the house.

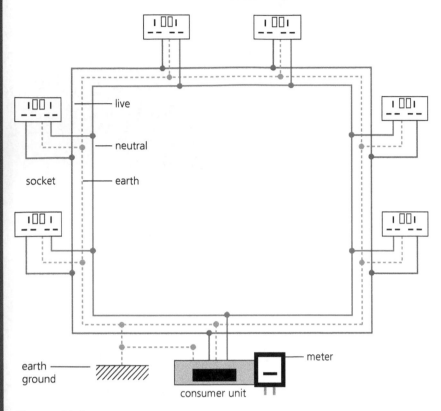

Figure 26.6 A domestic 32 A ring main.

The ring main cable consists of three wires: the live wire (always coloured brown), which carries the current from the consumer unit to the sockets; the neutral wire (always coloured blue), which returns the current to the consumer unit; and the earth wire (always coloured yellow/green), which acts as a safety system for the circuit. The earth wire connects the material of the whole installation to an earth connection at the local sub-station. If the live wire becomes loose and creates a short circuit, the electric current flows down the earth wire safely into the ground, reducing the risk of electrocution.

Now test yourself

TESTED

3 Calculate the current normally drawn from a 2.2 kW mains kettle, and use this value to decide on the rating of the fuse that should be fitted to its plug.
4 What is the difference between a.c. and d.c.?
5 Explain how an electricity consumer unit is used as a safety system for a house.
6 What is the difference between an mcb and a cartridge fuse?

Answers online

Micro-generation of domestic electricity

REVISED

Micro-generation of electricity means generating electricity locally on a small scale and close to where it is needed. Examples of micro-generation are roof-top photovoltaic cells and domestic wind turbines. Micro-generation has many advantages and some disadvantages over the large-scale generation of electricity from power stations.

Advantages of micro-generation

- Does not produce carbon dioxide and so does not contribute towards the greenhouse effect and global warming.
- Does not produce sulfur dioxide or oxides of nitrogen and so does not contribute towards acid rain.
- Zero fuel costs.
- Higher efficiency of generation.
- Can sell some electricity back to National Grid (feed-in).
- Roof-top photovoltaics:
  ○ provide 'free' electricity during daylight hours
  ○ on average, can generate 3 kW of electricity (peak).
- Domestic wind turbines:
  ○ provide 'free' electricity when the wind is blowing
  ○ on average, can generate 6 kW of electricity (peak).

Disadvantages of micro-generation

- Erratic energy supply.
- Can have long payback times.
- Cannot generate large quantities of electricity in one place.
- Many locations are very limited in which types of micro-generation can be used.
- Some people object to the visual impact of wind turbines and solar panels.
- Roof-top photovoltaics:
  ○ have a visual impact on roof-tops
  ○ need to cover a large area to generate large amounts of electricity.
- Domestic wind turbines:
  ○ cause a visual impact from the turbine
  ○ cause an impact from the noise of the turbine
  ○ are unsuitable for most locations as they need an exposed windy site.

Exam tip

A common extended writing question involves comparing micro-generation systems, such as wind turbines and roof-top photovoltaics, with more conventional methods of generating electricity, such as gas-fired power stations. Remember to compare the advantages and disadvantages of both methods.

Remember to use capital letters and full stops in your answer, and spell the keywords correctly.

Now test yourself

TESTED

7 A roof-top wind turbine costs £3500 to install and will save £700 of electricity costs per year. Calculate the payback time of the turbine.
8 Wind turbines and roof-top photovoltaics both produce 'free' energy. Why are photovoltaics less erratic than wind turbines?

Answers online

Summary

- The kilowatt (kW) is a convenient unit of power used by domestic appliances. The kilowatt-hour (kWh) is the unit of energy used by electricity companies when charging customers.
- The cost of electricity can be calculated using the equations:

  units used (kWh) = power (kW) × time (h)

  cost = units used × cost per unit

- The power of a domestic appliance can be obtained either directly from its power rating plate or through the energy banding (A–G) sticker.

- Electric currents can be either alternating current (a.c.) or direct current (d.c.). Direct currents only flow in one direction, whereas alternating current flows in one direction for half a cycle, then in the opposite direction for the rest of the cycle.
- Fuses, miniature circuit breakers (mcb) and residual current circuit breakers (rccb) are devices in mains electrical circuits (and appliances) to limit the current flowing through the circuit, making the circuits safer. The rating of a fuse is the maximum current that can flow through it before the special wire within the fuse

melts, disconnecting the circuit. The rating of the fuse fitted in a plug is always a little higher than the standard operating current of the appliance.
- A domestic ring main is a way of connecting up the sockets within a house. The live wire carries the electricity from the consumer unit, the neutral wire returns it to the consumer unit and the earth wire acts as a safety system in case there is a short circuit.

- The cost effectiveness of introducing domestic solar and wind energy equipment into a house is determined by the installation cost of the equipment and the fuel cost savings. The payback time is the amount of time (in years) that the equipment needs to be fitted for before the savings start to outweigh the installation costs.

Exam practice

1 Fuses and circuit breakers are electrical safety devices used to protect household electrical circuits.
 a) Explain how fuses and miniature circuit breakers protect household electrical circuits. [2]
 b) State one way in which miniature circuit breakers are more effective than fuses. [1]
 c) Explain how the action of a residual current device is different from that of a miniature circuit breaker. [2]

(WJEC GCSE Physics P2 Higher Tier Summer 2007 Q6)

2 Circuits and users are protected by the following safety features:
 fuse   miniature circuit breaker (mcb)   residual current circuit breaker (rccb)   earth wire
 a) Name a safety feature that prevents cables becoming too hot. [1]
 b) Name a safety feature that detects a difference in current between the live and neutral wires. [1]
 c) Name a safety feature that will cause a fuse to blow if current flows through it. [1]
 d) Name a safety feature that needs replacing once it acts. [1]

(WJEC GCSE Physics P1 Foundation Tier Summer 2008 Q1)

3 The lighting circuit in a house is protected by a 5A fuse and connected to 230V. The table shows the current taken by different lamps.

Power of lamp (W)	Current (A)
40	0.17
60	
100	0.43

 a) Use the equation below to find the current through a 60W lamp. [1]

$$current = \frac{power}{voltage}$$

 b) The circuit shows three lamps in a household lighting circuit connected to a 5A fuse. We can calculate the current through the fuse by adding up the currents through each of the lamps. Use the information in the table to find the current flowing through the fuse when all these lamps are switched on. [2]
 c) Find the maximum number of 100W lamps that could be connected in a 5A household lighting circuit. [2]

(WJEC GCSE Physics P2 Higher Tier January 2008 Q1)

4 The table shows information about three electrical appliances.

Appliance	Power (W)	Power (kW)	Units (kWh) used in 1 week
Kettle	2100	2.1	5
Electric oven		4.0	12
Microwave oven	900	0.9	1

 a) i) What does 'kW' stand for? [1]
  ii) Complete the table. [1]
  iii) State which appliance uses the most energy every second. [1]

b) Calculate the number of hours that the electric oven is used in 1 week. [1]

c) All three appliances are used for 1 week.
   i) Calculate the total number of units used. [1]
   ii) If 1 unit of electricity costs 12p. Calculate the cost of using all three appliances for 1 week. [2]

(WJEC GCSE Physics P1 Foundation Tier January 2011 Q7)

5 40% of all the wind energy in Europe blows over the UK, making it an ideal country for small home wind turbines. Roof-mounted turbines produce around 1 kW to 2 kW depending on wind speed. To be effective you need an average wind speed bigger than 5 m/s. Small domestic wind systems are particularly suitable for use in remote locations where homes are not connected to the National Grid. Costs for a roof-mounted wind system are £1500. Recent monitoring of a range of small domestic wind systems has shown that a well-sited 2 kW turbine could save around £300 a year off electricity bills.

a) Why is the UK ideal for small home wind turbines? [1]

b) The average wind speed in one town is 3.5 m/s. Give a reason why homeowners here would not be advised to install wind turbines. [1]

c) Why are wind turbines useful for supplying electricity to farms on hilltops well away from towns? [2]

d) Calculate the payback time for the roof-mounted wind turbine mentioned in the passage. [1]

(WJEC GCSE Physics P1 Foundation Tier Summer 2010 Q5)

Answers and quick quiz 26 online

ONLINE

27 Features of waves

Transverse and longitudinal waves

There are two types of waves: **transverse waves** (like water waves), in which the direction of motion of the wave is at right angles to the direction of vibration of the wave, and **longitudinal waves** (like sound waves), in which the direction of motion is in the same direction as the direction of vibration of the wave. Transverse waves travel as a series of peaks and troughs; longitudinal waves travel as compressions and rarefactions. Both of these sorts of waves can be demonstrated with a slinky spring as shown in Figure 27.1.

> A **transverse wave** is one in which the vibrations causing a wave are at right angles to the direction of energy transfer.
>
> A **longitudinal wave** is one in which the vibration causing the wave is parallel to the direction of energy transfer.

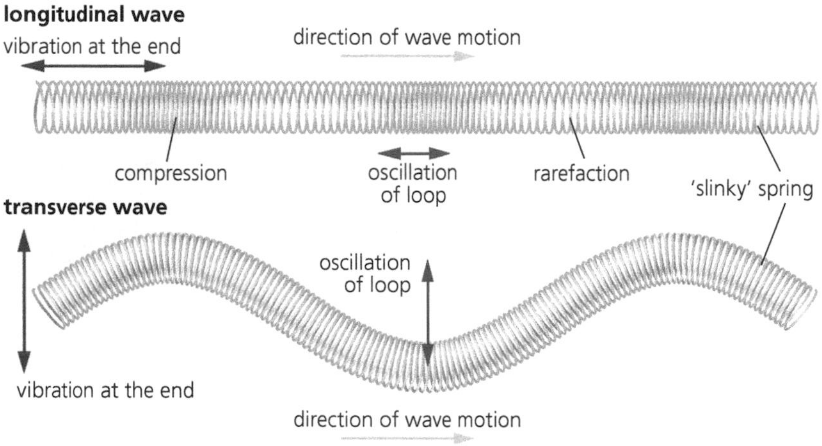

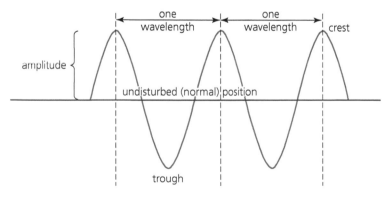

Figure 27.1 Longitudinal and transverse waves on a slinky.

How do we describe waves?

Waves are described in terms of their **wavelength**, **frequency**, **speed** and **amplitude**. Figure 27.2 shows these quantities on a transverse wave, like a water wave or light wave, in which the direction of vibration of the wave is at right angles to the direction of travel of the wave.

Figure 27.2 Wave measurements.

> The **frequency**, f, of a wave is the number of waves that pass a point in 1 second. Frequency is measured in hertz, Hz, where 1 Hz = 1 wave per second.
>
> The **speed** of a wave, v, is the distance that a wave travels in 1 second. Wave speed is measured in metres per second, m/s.
>
> The **amplitude** of a wave is a measure of the energy carried by the wave. Amplitude is measured from the undisturbed (normal) position to the top of a crest or the bottom of a trough. Loud sounds have larger amplitudes than quiet sounds.

> The **wavelength**, λ, of a wave is the distance that the wave takes to repeat itself – this is normally measured from one crest to the next crest. Wavelength is measured in metres, m.

Calculating wave speed, frequency and wavelength

- Wave speed can be calculated using the equation:

$$\text{wave speed} = \frac{\text{distance}}{\text{time}}$$

- Wave speed, frequency and wavelength are all related by the basic wave equation:

  wave speed = frequency × wavelength

$$v = f\lambda$$

- Waves travel at a range of different speeds.
- All electromagnetic waves travel at the speed of light, $c = 300\,000\,000$ m/s or 3×10^8 m/s.
- Water waves, like surf, travel at about 4 m/s.

Now test yourself

1 A surfer takes 10 s to travel 50 m on the crest of a wave onto a beach. What is her speed?
2 The wavelength of the waves in Question 1 is 40 m. What is the frequency of the waves?
3 Calculate the speed of sound waves travelling through wood with a frequency of 5 kHz and a wavelength of 79.2 cm.
4 Explain the difference between a transverse and a longitudinal wave.

Answers online

What is the electromagnetic spectrum?

The electromagnetic spectrum is a family of (transverse) waves that all travel at the same speed in a vacuum, $300\,000\,000$ m/s or 3×10^8 m/s. Like the energy given out by radioactive materials, electromagnetic waves are also called '**radiation**'.

> **Radiation** refers to electromagnetic waves or the energy given out by radioactive materials.

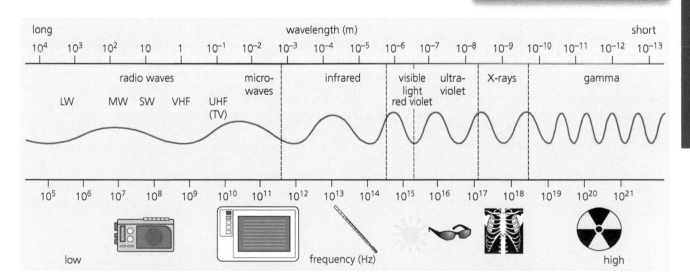

Figure 27.3 The electromagnetic spectrum.

The different parts of the electromagnetic spectrum have different wavelengths and frequencies, and energies. The higher the frequency of a wave, the higher its energy. The frequency, wavelength and energy of an electromagnetic wave completely determine its properties and how it will behave.

- Gamma rays have very high energy and can ionise (kill or damage) cancer cells, but they are also used to make images of the body.
- X-rays are also ionising, and are also used in medical imaging.
- Ultraviolet light can ionise skin cells causing sunburn.

The short wavelength parts of the electromagnetic spectrum (ultraviolet, X-rays and gamma rays) are all called ionising radiations, because they are able to interact with atoms and damage cells due to their large energies.

- Infrared radiation is used for heating and communications, such as in TV remote controls and optic fibres.
- Microwaves are also used for heating and communicating, particularly as mobile phone signals.
- Radio waves are used for communications over much longer distances, transmitting TV and radio programmes.

All parts of the electromagnetic spectrum can carry information and energy. Stars also emit all parts of the spectrum, giving us information about their composition and behaviour. Visible light, infrared, microwaves and radio waves are commonly used by human beings to transmit information.

> **Exam tip**
>
> Electromagnetic spectrum diagrams can be drawn in either direction. Remember: radio waves have the longest wavelengths, lowest frequencies and lowest energies. Gamma rays have the shortest wavelengths, highest frequencies and highest energies. You must learn the order.

Now test yourself

TESTED

5 What is meant by ionising radiation?
6 Which part(s) of the electromagnetic spectrum:
  a) has the lowest energy
  b) has a frequency range between visible light and X-rays
  c) can be used for cooking
  d) are emitted by stars
  e) are used for human communications?

Answers online

Reflection of waves

REVISED

Reflection is a fundamental property of all waves. When straight (plane) wave-fronts hit a flat barrier, they rebound off, obeying the law of reflection. Figure 27.4 shows this happening with water waves in a ripple tank.

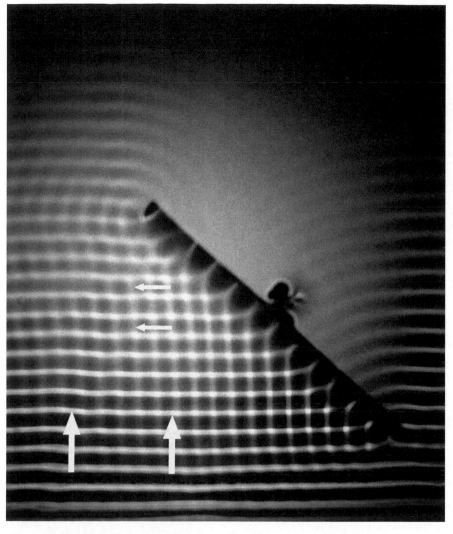

Figure 27.4 Water waves reflecting off a plane barrier in a ripple tank.

Figure 27.5 explains the reflection of waves.

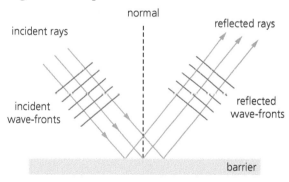

Figure 27.5

> The **normal** is a line drawn at 90° to a surface where waves are incident.
>
> The **angle of incidence** is the angle between the incident ray and the normal.
>
> The **angle of reflection** is the angle between the reflected ray and the normal.

The imaginary rays, drawn at right angles to the wave-fronts, show the direction of travel of the wave-fronts. The angles between the incident and reflected rays and the **normal** line (an imaginary line at right angles to the barrier/mirror) are equal, obeying the law of reflection, where:

angle of incidence = angle of reflection

Refraction of waves

REVISED

When water waves travel from deep water into shallow water, they slow down and the wave-fronts get closer together, decreasing their wavelength. This effect is called **refraction**. When the wave-fronts hit the boundary between the deeper water and the shallow water at an angle, they appear to change direction as shown in Figure 27.6.

Figure 27.6 The refraction of water waves in a ripple tank.

The refraction of light through a glass block in Figure 27.7 shows the rays of light changing direction as they go from the air into the glass, and then back again.

Figure 27.8 is a diagrammatic version of Figure 27.7, with the angles labelled.

Figure 27.7 The refraction of light through a glass block.

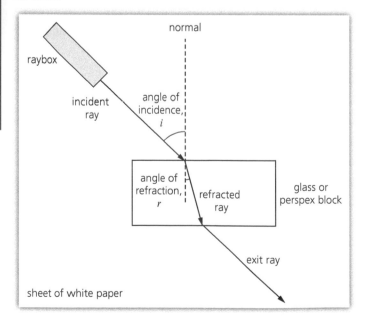

Figure 27.8 The angle of incidence and the angle of refraction.

Communicating using satellites

Mobile phones use microwaves and, as they are wireless signals, you don't need a copper cable or an optical fibre to transmit them. However, one disadvantage of using microwaves is that there must be a clear path between the transmitter and the receiver, which might be your television aerial or mobile phone. To cover the largest area, TV and mobile phone transmitters are tall and sited on hills. The curvature of the Earth means that repeater stations have to relay the microwave signal to distant transmitters. Satellites must be used for long-distance communications around the world. Theoretically, only three satellites are needed to transmit signals around the world. In practice, more are used. The satellites are placed in orbit at a height of 36 000 km. They orbit the Earth, above the equator, exactly in time with the Earth's rotation. This is called a geosynchronous (geostationary) orbit.

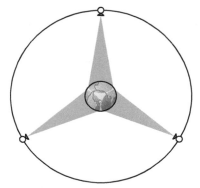

Figure 27.9 The Earth from above the North Pole; three geosynchronous satellites could send signals to the Earth.

Now test yourself

7 What is the law of reflection?
8 What is the 'normal' line?
9 Explain why the wavelength of water waves decrease as they travel from deep water to shallow water.
10 Why are three satellites needed for long-distance microwave communications around the world?

Answers online

Summary

- Transverse waves vibrate at right angles to their direction of motion of the wave. Longitudinal waves vibrate in the same direction as the direction of motion.
- Waves can be distinguished in terms of their wavelength, frequency, speed, amplitude (and energy).
- The equations associated with waves are:

$$\text{wave speed} = \frac{\text{distance}}{\text{time}}$$

wave speed (m/s) = frequency (Hz) × wavelength (m)

- Waves will reflect when they hit a barrier, obeying the law of reflection.
- Waves will refract when they go across a boundary between one medium where they travel at one speed into another medium where they travel at a different speed. Refraction causes the wavelength of the wave to change.

- All regions of the electromagnetic spectrum transmit information and energy.
- The electromagnetic spectrum is a continuous spectrum of waves of different wavelengths and frequencies consisting of radio waves, microwaves, infrared, visible light, ultraviolet radiation, X-rays and gamma rays, but all the waves travel at the same speed in a vacuum – the speed of light.
- The term 'radiation' can be used to describe both electromagnetic waves and the energy given out by radioactive materials.
- Radioactive emissions and the short wavelength parts of the electromagnetic spectrum (ultraviolet, X-ray and gamma ray) are ionising radiations, and they can interact with atoms, damaging cells by the energy that they carry.
- Microwaves and infrared radiation are used for mobile phones, intercontinental optical-fibre links, and for long-distance communication, via geosynchronous satellites.

Exam practice

1 The graph in Figure 27.10 shows how the frequency of deep ocean waves depends on the wavelength of the waves.

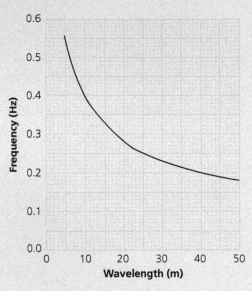

Figure 27.10

a) Use information from the graph in Figure 27.10 and the equation below to calculate the speed of waves with a wavelength of 40 m.   [2]

wave speed = wavelength × frequency

b) A large meteorite falls into the ocean and produces waves with a range of wavelengths.
    i)   Use the equation below to calculate how long it would take 40 m-wavelength waves to arrive at an island 5600 m away.   [1]

$$\text{speed} = \frac{\text{distance}}{\text{time}}$$

    ii) Would 10 m waves arrive before or after the 40 m waves? Use information from the graph to explain your answer.   [2]

(WJEC GCSE Physics P1 Foundation Tier Summer 2009 Q7)

2 Figure 27.11 shows a train of waves.

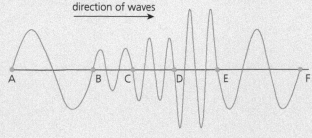

Figure 27.11

a) How many waves are shown between A and C? [1]

b) Between which two of the points, A–F, is:

   i) the wavelength biggest [1]

   ii) the amplitude smallest? [1]

c) The eight waves between A and F cover a distance of 240 cm. Calculate the average wavelength of the waves. [1]

(WJEC GCSE Physics P1 Higher Tier January 2011 Q1)

3 Yellow light travels to us from the Sun at a speed of 3×10^8 m/s. It has a frequency of 5×10^{14} Hz. Write down in words a suitable equation and use it to calculate the wavelength of this yellow light. [3]

(WJEC GCSE Physics P1 Higher Tier Summer 2008 Q5(a))

4 a) Using the words below fill in the missing parts of the electromagnetic spectrum, (i) and (ii). [2]

ultraviolet   radio waves   sound waves   water waves

(i)	Microwaves	Infrared	Visible light	(ii)	X-rays	Gamma rays

b) Some electromagnetic waves can be used for communications.

   i) Name the wave that is used by remote controls. [1]

   ii) Name the wave that is used to communicate with satellites in space. [1]

c) Some of these waves can be harmful.

   i) Name one wave from the list that can ionise cells in the body. [1]

   ii) What is the danger from a large dose of infrared rays? [1]

(WJEC GCSE Physics P1 Foundation Tier January 2011 Q6)

5 Figure 27.12 shows a communications satellite A in geosynchronous (geostationary) orbit around the Earth. The diagram is not to scale.

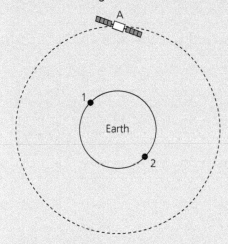

Figure 27.12

a) **i)** Explain the advantages of placing communications satellites in geosynchronous orbit. [2]

   ii) Copy the diagram and add another satellite B and repeater station 3 that will enable radio station 1 to communicate with radio station 2. [2]

   iii) Show on the diagram the path taken by the signal, via the satellites A and B, when radio station 1 communicates with radio station 2. [1]

b) **i)** Communications between geosynchronous satellites and Earth are made using microwaves of wavelength 20 cm that travel at 3×10^8 m/s. Use a suitable equation to calculate the frequency of the microwaves. [3]

   ii) The time delay between sending a signal from 1 and its reception at 2 is 0.48 s. Use a suitable equation to find the approximate height of geostationary satellites above the Earth. [3]

(WJEC GCSE Physics P1 Higher Tier January 2010 Q6)

→

6 a) Electromagnetic waves are used in communications to send television signals.
   i) Name the part of the spectrum that carries TV signals via satellites. [1]
   ii) Name the part of the spectrum that carries TV signals from transmitters to a home TV aerial. [1]
   iii) Name a part of the spectrum that carries TV signals through optical-fibre cables. [1]

b) A householder installs a dish to receive TV signals from a communication satellite. Explain why the householder will not need to move the dish once it is set up. [2]

(WJEC GCSE Physics P1 Foundation Tier Summer 2010 Q8)

7 Figure 27.13 shows light travelling from air into a glass block.

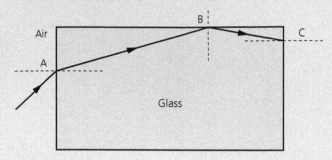

Figure 27.13

a) i) What name is given to the bending of light at point A? [1]
   ii) Give a reason why the light changes direction at A. [1]
b) State what law is being obeyed at point B. [1]
c) At point C, the light passes out into the air.
   i) Give one reason why it does not go back into the block as it does at point B. [1]
   ii) Draw the ray direction into the air at point C. [1]
d) A long and very thin glass block becomes an optical fibre. Name a type of the electromagnetic radiation (other than visible light) that can be used to send messages along an optical fibre. [1]
e) The speed of signals along optical fibres is 2.0×10^8 m/s. Select an equation and use it to find the time that a signal would take to travel from London to New York along an optical fibre if the distance is 4.8×10^7 m. Give the correct unit for your answer. [3]

(WJEC GCSE Physics P1 Higher Tier January 2011 Q3)

Answers and quick quiz 27 online

ONLINE

28 Distance, speed and acceleration

Describing motion REVISED

The motion of an object can be described using the following quantities:
- distance (measured in metres, m): how far the object travels, or how far away the object is from a fixed point
- time (measured in seconds, s): the time interval between two events or the time since the start of the motion
- speed (measured in metres per second, m/s): a measure of how fast or slow the object is moving. The speed of the object can be calculated using the equation:

$$\text{speed} = \frac{\text{distance}}{\text{time}}$$

- velocity (measured in metres per second, m/s, in a given direction): a measure of how fast or slow the object is moving in a given direction (e.g. left/right, North/South), which is the speed in a given direction.
- **acceleration** or deceleration (measured in metres per second per second, m/s²): the rate that the object is speeding up or slowing down, which is the rate of change of velocity. Acceleration can be calculated using the equation:

$$\text{acceleration or deceleration} = \frac{\text{change in velocity}}{\text{time}}$$

- Speed is a scalar quantity because it only has magnitude (size); velocity is a vector quantity because it has direction as well as magnitude.

> **Acceleration** is the rate of change of velocity.

Now test yourself TESTED

1 Calculate the speed of a horse that gallops 200 m in 16 s.
2 Calculate the acceleration of the horse if it takes 5 s to get from rest (0 m/s) to galloping at 12.5 m/s.

Answers online

Graphs of motion REVISED

The motion of objects can be described and analysed using graphs of motion. There are two types of motion graph: distance–time graphs and velocity–time graphs.

Distance–time graphs

- A distance–time graph allows us to measure the speed of a moving object.
- The graph in Figure 28.1 shows an object moving away from a starting point at a constant speed of 6 m/s.
- Stationary objects are represented by straight horizontal lines.
- The slope or gradient of a distance–time graph is the speed of the object.

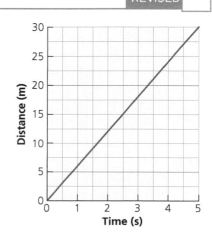

Figure 28.1 Distance-time graph.

Velocity–time graphs

- A velocity–time graph gives us more information than a distance–time graph. The graph in Figure 28.2 shows an object that is:
  - stationary for 2 seconds
  - accelerating at $3\,\text{m/s}^2$ for 2 seconds
  - moving at a constant velocity of $6\,\text{m/s}$ for $6\,\text{s}$.
- The slope or gradient of a velocity–time graph is the acceleration of the object.
- The distance travelled by the object is the area under the velocity–time graph (in this case $42\,\text{m}$).

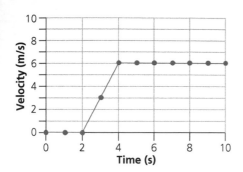

Figure 28.2 Velocity–time graph.

Exam tip

You need to be very careful when you are analysing and extracting information from a graph in a question. Many students make mistakes by reading the axes incorrectly. A good technique to help you measure quantities from graphs is to use a sharp pencil and a ruler to draw thin guidelines onto the axes at the places where you need to take a reading.

Now test yourself

TESTED

3 Sketch distance–time graphs for the following motions:
   a) an object moving at a constant speed of $6\,\text{m/s}$ for 5 seconds
   b) an object stationary, $20\,\text{m}$ away from an observer, for $5\,\text{s}$
   c) an object moving at a constant speed of $10\,\text{m/s}$ for 20 seconds, then $5\,\text{m/s}$ for $20\,\text{s}$.
4 Sketch velocity–time graphs of the following motions:
   a) an object stationary for $2\,\text{s}$, then accelerating at $3\,\text{m/s}^2$ for $2\,\text{s}$, then travelling at a constant speed of $6\,\text{m/s}$ for $6\,\text{s}$
   b) an object initially travelling at $9\,\text{m/s}$ and decelerating at $3\,\text{m/s}^2$ for $3\,\text{s}$, then accelerating at $2\,\text{m/s}^2$ for $4\,\text{s}$, then travelling at a constant velocity of $8\,\text{m/s}$ for $3\,\text{s}$
   c) an object initially stationary, accelerates at $3\,\text{m/s}^2$ for $3\,\text{s}$, then travels at a constant velocity of $9\,\text{m/s}$ for $4\,\text{s}$, then decelerates $3\,\text{m/s}^2$ for $3\,\text{s}$ back to stationary
   d) an object initially travelling at $6\,\text{m/s}$ for $3\,\text{s}$, then accelerates at $2\,\text{m/s}^2$ for $1\,\text{s}$, then remains at $8\,\text{m/s}$ for $1\,\text{s}$ before decelerating at $2\,\text{m/s}^2$ for $4\,\text{s}$ to stationary.

Answers online

Stopping distances

REVISED

Vehicles do not stop instantaneously – there is a time delay between the driver seeing the need to stop, such as a potential hazard, and the vehicle stopping. During this time, the vehicle is still travelling at speed, so it travels through a distance. The total stopping distance of a vehicle is made up of the 'thinking distance' and the 'braking distance'.

- Thinking distance is the distance the vehicle travels while the driver sees the hazard, thinks about braking and then actually reacts by braking.
- Braking distance is the distance that the vehicle moves while the brakes are being applied and the vehicle is decelerating to $0\,\text{m/s}$.

total stopping distance = thinking distance + braking distance

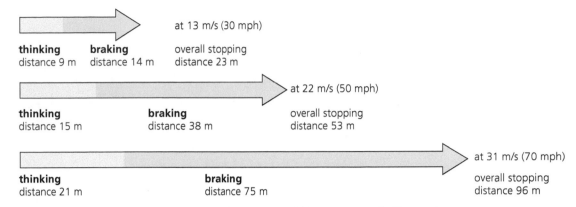

Figure 28.3 Thinking and braking distance at different speeds (from the Highway Code).

Factors affecting stopping distances

Thinking distance depends on several different factors, including:
- the velocity of the car
- the reaction time of the driver (which depends on tiredness, alcohol use, etc.)
- whether the driver is distracted, such as by hearing a mobile phone ring.

The braking distance also depends on several factors:
- the velocity of the car
- the mass of the car
- the condition of the brakes
- the condition of the tyres
- the condition of the road surface
- the weather.

Stopping safely

When a car stops very quickly, for example in a collision, in order to minimise injuries to the occupants, the key factor is to reduce the forces that act on them. Car manufacturers have built safety systems into modern cars to reduce these forces: seat belts, air bags and crumple zones.

According to Newton's second law:

$$\text{resultant force, } F\ (\text{N}) = \frac{\text{change in momentum, } \Delta p\ (\text{kg m/s})}{\text{time for change, } t\ (\text{s})}$$

$$F = \frac{\Delta p}{t}$$

There are two ways of reducing the force on the occupants:
1 by reducing the speed of the collision and so reducing the change of momentum
2 by increasing the time for the collision.

All three safety systems listed above work by increasing the time of the collision – by allowing something to be deformed during the collision. Seat belts stretch; airbags slowly deflate; crumple zones crumple in on themselves.

Now test yourself

5 What is the difference between thinking distance and braking distance?
6 List three factors that affect thinking distance.
7 How do seat belts reduce the force of an impact on a driver?

Answers online

Traffic control measures

The speed of traffic along a road can be controlled by imposing speed limits. Urban areas typically have speed limits of 30 mph, while the national speed limit on single carriageway roads is 60 mph and 70 mph on dual carriageways and motorways. To slow down traffic even more, particularly around schools and residential areas, speed bumps can be installed, forcing drivers to slow down as they drive over the bumps.

Summary

- Speed is a measure of how fast an object is moving:

$$\text{speed} = \frac{\text{distance}}{\text{time}}$$

- Speed is a scalar quantity because it only has magnitude (size).
- Velocity is a vector and has direction as well as magnitude.
- Velocity is measured in metres per second, in a specified direction.
- Acceleration is the rate of change of velocity. Objects that are getting faster are accelerating, and objects that are getting slower are decelerating:

$$\text{acceleration or deceleration} = \frac{\text{change in velocity}}{\text{time}}$$

- The units of acceleration are metres per second squared, m/s^2.
- On distance–time graphs stationary objects are shown by flat, straight lines and objects

travelling at constant velocity are shown by sloping, straight lines.
- The speed of an object can be found by measuring the slope or gradient of the graph.
- For velocity–time graphs, a flat, straight line indicates an object travelling at constant velocity and a straight line sloping upwards indicates an object accelerating. A straight line sloping downwards indicates an object decelerating.
- The acceleration can be found by measuring the gradient or slope of the velocity–time graph.
- **H** The area under the velocity–time graph is the distance travelled.
- The safe stopping distance of a vehicle depends on the driver's reaction time (which affects the thinking distance) and the braking distance of the vehicle.
- Traffic control measures include speed limits and speed bumps.

Exam practice

1 During road tests, three cars are tested to find out how long they take to accelerate from 0 to 60 mph (27 m/s). The results are shown in the table.

Car	Time to reach 60 mph from rest (s)
W	5
X	8
Y	9

a) State which car, W, X or Y, has the smallest acceleration. [1]
b) A velocity of 60 mph is the same as a velocity of 27 m/s. Select a suitable equation and use it to calculate the acceleration of car Y during the test in m/s^2. [3]

(WJEC GCSE Physics P3 Foundation Tier Summer 2009 Q2)

→

2 A theme park ride involves a group of people being lifted in a carriage and then dropped from a height. The graph in Figure 28.4 shows the motion of such a ride.

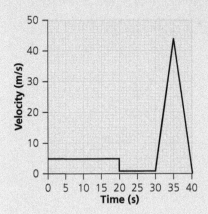

Figure 28.4

a) Describe the motion of the carriage in the first 20 s. [1]
b) Select a suitable equation and use it to find the acceleration of the carriage between 30 s and 35 s. [2]

(WJEC GCSE Physics P2 Higher Tier Summer 2009 Q3)

3 A car overtakes a lorry. In doing so, the car accelerates and, after overtaking safely, returns to its original speed. The graph in Figure 28.5 represents the motion of the car when overtaking the lorry.

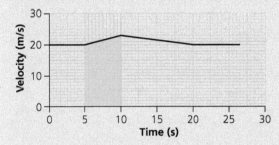

Figure 28.5

a) Select a suitable equation and then use it, together with data from the graph, to calculate the acceleration of the car during overtaking. [4]
b) Describe clearly what the shaded area of the graph represents. [2]
c) Use the data from the graph to calculate the distance travelled between 10 s and 20 s. [3]

(WJEC GCSE Physics P3 Higher Tier Summer 2008 Q2)

4 The overall stopping distance of a car is made up of two parts: thinking distance and braking distance. At a speed of 20 m/s the Highway Code states that a car has a thinking distance of 12 m and a braking distance of 40 m.
a) Use a suitable equation to find the thinking time for a driver. [2]
b) Complete the table below. Some boxes have been completed for you. [3]

Condition	Effect on thinking distance	Effect on braking distance	Effect on overall stopping distance
Poor brakes	No change	Increases	Increases
Driver under the influence of alcohol			Increases
Driver drives at a lower speed	Decreases		
Wet road		Increases	

(WJEC GCSE Physics P2 Foundation Tier Summer 2010 Q5)

Answers and quick quiz 28 online

ONLINE

29 Newton's laws

Inertia and Newton's first law of motion

REVISED

The mass of an object dictates how easy (or difficult) it is to get the object moving or to change its motion. This property is called **inertia**, defined as the resistance of any object to a change in its state of motion or rest. Very massive objects, such as the International Space Station, have very large inertias. Altering their motion takes a very large force.

> **Inertia** is the opposition of an object to a change in its motion.

Newton's first law

In 1687 Isaac Newton realised the link between the motion of an object and the force on it. He summarised this in his first law of motion. On Earth it is very difficult to observe Newton's first law, because friction always acts to oppose the motion of an object.

> 'An object at rest stays at rest, or an object in motion stays in motion with the same speed and in the same direction, unless acted on by an unbalanced force.'

Resultant forces and Newton's second law of motion

REVISED

When several forces act on an object at the same time, they either cancel each other out (balanced forces), or they combine together to produce a resultant (unbalanced) force. Figure 29.1 shows two forces acting on a lorry. The two forces combine to produce a single resultant force of 800 N in the direction of motion.

- A resultant force acting on an object causes a change in its motion. (The lorry will accelerate.)
- Balanced forces cause an object to remain stationary or move at constant speed.

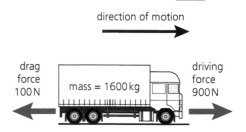

Figure 29.1 Unbalanced forces.

Experiments show us that, as the resultant force on an object increases, so does the size of the acceleration. If we double the resultant force, then the acceleration doubles. The resultant force and the acceleration are proportional to each other. When a graph of resultant force against acceleration is analysed further, we find that the gradient of the line is equal to the mass of the object.

Expressing this as a word equation, we can say:

resultant force, F (N) = mass, m (kg) × acceleration, a (m/s^2)

$$F = ma$$

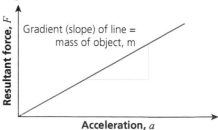

Figure 29.2 A graph of a resultant force against acceleration is a straight line.

Now test yourself

TESTED

1 A lorry, of mass 1600 kg, starts at rest and accelerates to 20 m/s in 40 s. Calculate its:
   a) acceleration
   b) resultant force.
2 State:
   a) Newton's first law of motion
   b) Newton's second law of motion.

Answers online

Falling objects

The mass of an object is a measure of how much matter (stuff) there is in the object. Mass, *m*, is measured in kilograms, kg. The weight of an object is the force of gravity acting on the object's mass. Weight is measured in newtons, N. On the surface of the Earth the weight of a 1 kg object is approximately 10 N. The weight of any object on the Earth's surface can be calculated by multiplying its mass, in kg, by 10. The weight of a 1600 kg lorry is therefore 16 000 N.

weight (N) = mass (kg) × gravitational field strength (N/kg)

As an object, such as a parachutist, falls, the weight remains constant. Initially, the only (resultant) force on the parachutist is her weight, so she accelerates downwards. As she speeds up, the force of air resistance acting upwards on her increases. Eventually, it is equal and opposite to her weight – the two forces are equal in size but acting in opposite directions, so they are balanced. The parachutist continues to fall, but at a constant (terminal) speed.

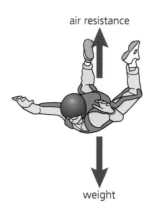

air resistance

weight

Figure 29.3 Balanced forces mean a moving object travels at constant speed.

Now test yourself

3 Calculate the weight on Earth of an 80 kg skydiver, if *g* = 10 N/kg.
4 Explain why a skydiver will accelerate when he initially jumps out of an aircraft.
5 Explain why a skydiver will eventually fall at a terminal speed.

Answers online

Interaction pairs

Forces between two objects always act in pairs. The force on one object is called the action force; the force on the other object is called the reaction force. Together they form an interaction pair. The action force and the reaction force are equal and opposite. But they do not cancel each other out because they act on different objects. This is known as Newton's third law:

'For every action force, there is an equal and opposite reaction force.'

For example, in Figure 29.4 some rugby players are practising scrummaging against a static scrum-machine and are pushing with a combined force of 500 N (the action force). Newton's third law means that the scrum-machine exerts a reaction force on the players of 500 N in the opposite direction.

action force of players
on machine

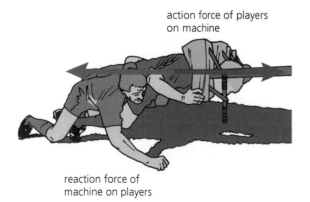

reaction force of
machine on players

Figure 29.4 The action and reaction forces are equal and opposite.

Some forces (like those involved in scrummaging) are contact forces: the two objects must come into contact with each other to exert the force. Other forces are 'action-at-a-distance' forces, such as gravity, or the forces exerted by electric and magnetic fields. When considering interaction pairs you need to remember that:

1 The two forces in the pair act on different objects.
2 The two forces are equal in size, but act in opposite directions.
3 The two forces are always the same type, for examⅢⅢple contact forces or gravitational forces.

Now test yourself
TESTED

6 State Newton's third law of motion.
7 During a rugby tackle, a tackler exerts a force of 200 N on a tackled player.
   a) What is the size of the force that the tackled player exerts on the tackler?
   b) In which direction does the force in (a) act?
   c) What type of forces are those in this question?

Answers online

Exam tip

When a diagram is given that shows several interaction pairs of forces, it is a good idea to highlight the individual pairs. This makes it easier to analyse them under the pressure of the exam.

Summary

- The mass of an object affects how easy or difficult it is to change the movement of that object. Massive bodies have large amounts of inertia, so require a large force to change their motion, or to make them move if they are stationary.
- Newton's first law of motion states that an object at rest stays at rest or an object in motion stays in motion with the same speed and in the same direction unless acted on by an unbalanced force.
- Newton's second law of motion states that: force = mass × acceleration; that is, the acceleration of an object is directly proportional to the resultant force and inversely proportional to the object's mass.
- Weight is the force of gravity acting on the mass of an object.
- On the surface of Earth, 1 kg of mass has a weight of 10 N; this is called the gravitational field strength, g.

- When an object is falling through the air, initially it accelerates and increases in speed because of the force of gravity acting on it. However, then the force of friction (air resistance) increases, causing the object to decelerate. Eventually, the air resistance force equals the weight of the object and it is said to be falling at its terminal (constant) velocity.
- Newton's third law states: In an interaction between two objects, A and B, the force exerted by body A on body B is equal and oppositely directed to the force exerted by body B on body A.
- Together the action force and the reaction force make an interaction pair.
- Forces may be 'contact' forces, where objects need to come into contact with each other to exert the force, or they may be 'action-at-a-distance' forces, such as gravity or electromagnetic forces.

Exam practice

Exam tip

Questions involving selection of answers from a list, as in Question 1, may appear straightforward. But they can easily catch people out. Read the question really carefully – at least twice, and make sure that the answer you are selecting actually addresses the question being asked. A good technique is to highlight the key words in the question, for example 'speeds up' and 'air resistance' in Question 1 (a).

1 The diagram in Figure 29.3 shows two forces acting on a skydiver. Choose the correct phrase in each set of brackets in the following sentences.
   a) When the skydiver speeds up, the air resistance is [bigger than / equal to / smaller than] the weight.   [1]
   b) When the skydiver falls at the terminal speed, the air resistance is [bigger than / equal to / smaller than] the weight.   [1]
   c) When the parachute is opened, the air resistance [gets bigger / stays the same / gets smaller] and the skydiver [goes back up / stays in the same place / continues to fall].   [2]

(WJEC GCSE Physics P2 Foundation Tier January 2009 Q2)

2 Figure 29.5 shows a test rocket on its launch pad. The rocket is powered by three engines, each of which produces a thrust of 2000 N. The mass of the rocket and its fuel is 500 kg, so that its weight is 5000 N.

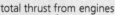

total thrust from engines

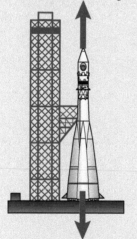

weight of rocket plus fuel

Figure 29.5

   a) When the engines are fired:
      i)   calculate the total thrust on the rocket   [1]
      ii) explain why the rocket moves upwards   [1]
      iii) calculate the resultant force on the rocket.   [1]
      iv) Select a suitable equation and use it to calculate the take-off acceleration of the rocket. [3]
   b) After 2 s, the rocket engines have used up 20 kg of fuel. Assuming that the thrust of the engines is constant, calculate:
      i)   the mass of the rocket and fuel after 2 s   [1]
      ii) the resultant force in newtons on the rocket after 2 s   [1]
      iii) the acceleration of the rocket after 2 s.   [1]
   c) Assuming that the thrust of the engines is constant, explain why the acceleration of the rocket will continue to increase for as long as the engines are fired.   [2]

(WJEC GCSE Physics P2 Higher Tier January 2010 Q4)

3 A skydiver of mass 60 kg weighs 600 N.
   a) The list on the left gives statements about the forces acting on a skydiver falling through the air. The list on the right gives five possible effects of these forces on the motion of the skydiver. Draw one line from each box on the left to the correct box on the right.   [3]

→

The air resistance is greater than the weight.	The skydiver slows down.
The air resistance is equal to the weight.	The skydiver moves upwards.
The weight is greater than the air resistance.	The skydiver speeds up.
	The skydiver falls at constant speed.
	The skydiver stops.

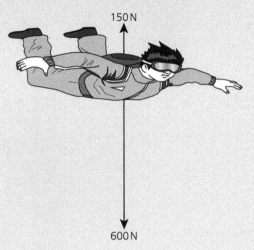

Figure 29.6

b) Figure 29.6 shows the forces acting on the skydiver at one point in her fall.

   i)   Calculate the resultant force acting on the skydiver.   [1]

   ii)   Select and use a suitable equation to calculate the acceleration produced by this resultant force.   [2]

c) The graph in Figure 29.7 shows how the speed changes with time for the skydiver. Choose letters from the graph which complete the following questions.

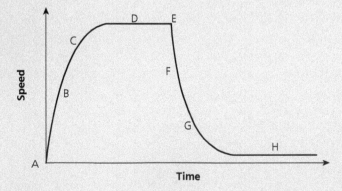

Figure 29.7

   i)   The point at which the skydiver opens the parachute is .....   [1]

   ii)   The skydiver is at terminal speed with the parachute open at .....   [1]

   iii)   Describe and explain the motion of the skydiver in terms of forces. Include in your answer:
- the type of motion in each region, AB, BC, and CD
- the forces acting in each region and how they compare.   [6 QWC]

(WJEC GCSE Additional Science/Physics P2 Foundation Tier May 2016 Q1 and Higher Tier Q6(b))

Answers and quick quiz 29 online

ONLINE

30 Work and energy

Work

When a force causes an object to move, or acts on a moving object, energy is transferred. The force moves through a distance and energy is transferred as **work**, measured in joules, J.

> **Work** is a measure of the energy transferred.

The amount of work done is calculated by:

work = force × distance moved in the **direction of the force**

$$W = Fd$$

For example, in rugby, the players lifting the jumper in a lineout exert an upward force, moving the jumper through a distance.

If the lifting force is 1000 N and the player is lifted through 1.5 m then the work done is:

$$W = Fd = 1000 \times 1.5 = 1500 \text{ J}$$

The work done is a measure of the energy transferred. But the work done only equals the total energy transferred if no energy is lost as heat to the surroundings (by air resistance or friction).

Figure 30.1 The force, *F*, moves through a distance, *d*.

Gravitational potential energy

When an object such as a ball is thrown or kicked vertically, the mass of the ball, *m*, is moved against the Earth's gravitational field strength, *g*, through a change in height, *h*, and it gains gravitational potential energy, PE.

$$\text{gravitational potential energy (PE)} = \text{mass, } m \text{ (kg)} \times \frac{\text{gravitational field}}{\text{strength, } g \text{ (N/kg)}} \times \frac{\text{change in}}{\text{height, } h \text{ (m)}}$$

$$PE = mgh$$

Now test yourself

1 Calculate the gravitational potential energy of a 0.44 kg rugby ball kicked vertically upwards to a height of 20 m. The gravitational field strength *g* = 10 N/kg.
2 A rugby player of mass 100 kg is lifted in a lineout, gaining 1500 J of gravitational potential energy. Calculate the height he is lifted. The gravitational field strength *g* = 10 N/kg

Answers online

Kinetic energy

When players run with the ball, their muscles convert chemical energy from their food into kinetic energy, KE (movement energy). We can calculate the kinetic energy of any moving object using the equation:

$$\text{kinetic energy (J)} = \frac{1}{2} \times \text{mass, } m \text{ (kg)} \times (\text{velocity, } v)^2 \text{ (m/s)}^2$$

$$KE = \frac{1}{2}mv^2$$

> **Exam tip**
>
> The equation for kinetic energy is unusual in that it involves the square of the velocity, v^2. A common mistake is to forget to square the velocity. Square the velocity first, then multiply by the mass and 0.5.

TESTED

Now test yourself

3 A rugby player can run with a rugby ball with a mean velocity of about 10 m/s. He has a mass of 80 kg. When running at 10 m/s, what is his kinetic energy?

4 A rugby ball of mass 0.44 kg is passed from one player to another with a kinetic energy of 2 J. Calculate the mean velocity of the ball.

Answers online

Total energy

REVISED

When objects such as rugby balls move, there is an interplay of gravitational potential energy and kinetic energy. The total energy of the ball stays constant, assuming that there is no energy lost through air resistance or friction.

total energy = gravitational potential energy + kinetic energy

total energy = PE + KE

Figure 30.2 shows the energy transformations during the flight of a ball after kicking upwards.

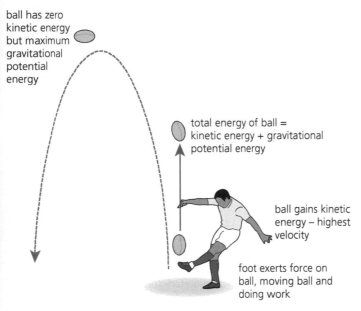

ball has zero kinetic energy but maximum gravitational potential energy

total energy of ball = kinetic energy + gravitational potential energy

ball gains kinetic energy – highest velocity

foot exerts force on ball, moving ball and doing work

Figure 30.2 The interaction of kinetic energy and gravitational potential energy on kicking a ball up.

Exam tip

Questions involving the interaction of gravitational potential energy and kinetic energy often come up in exams. Typically, they involve fairground rides, skiers, or bicycles going down slopes. You should try as many of these types of question as you can, to practise exchanging the two types of energy and doing the calculations.

Storing energy in springs

REVISED

When forces act on springs they can extend (get longer) or compress (get shorter). The spring extension (or compression) depends on the stiffness of the spring (through a value called the spring constant, k) and the force involved. The force, F (in N), spring constant, k (in N/m), and extension (or compression), x (in m), are related to each other using the equation:

force, F = spring constant, k × extension, x

$F = kx$

Very stiff springs require a lot of force to extend (or compress) them and they have very high spring constants. A graph of force against extension for a spring, obeying $F = kx$, is shown in Figure 30.3. The gradient (or slope) of the graph line is the spring constant, k, of the spring.

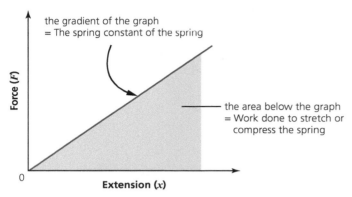

Figure 30.3 A force–extension graph for a spring.

When a force is exerted on a spring, it does work on the spring, because work is done when a force moves through a distance (the distance being the extension – or compression – of the spring). The work done stretching (or compressing) the spring can be measured by finding the area under the force–extension graph (a triangular shape). Where the spring obeys the equation $F = kx$, then the work done, W, is given by:

$$W = \frac{1}{2}Fx$$

Now test yourself

TESTED

5 A spring with a spring constant of 25 N/m is extended by 0.14 m. Calculate the force acting on the spring.
6 Explain how a force–extension graph can be used to determine the work done stretching a spring.
7 Calculate the work done stretching the spring in Question 5.

Answers online

Work, energy and vehicles

REVISED

Improving vehicle efficiency

The efficiency of vehicles can be maximised by:
- Improving their aerodynamics, by allowing the air to move more smoothly over the external surfaces of the vehicle. This greatly decreases the amount of work that the vehicle has to do forcing the air away, improving fuel economy and increasing the range of the vehicle.
- Lowering the bottom of the vehicle, making the wheels more enclosed by the wheel arches. This also improves the aerodynamics of the vehicle and reduces the air resistance, improving the fuel economy.
- Designing tyres to ensure a balance between the need for grip (to ensure the safety of the passengers) and the need to minimise the rolling resistance between the tyre and the road surface (also improving the fuel economy).
- Reducing the amount of energy lost when the vehicle is idling in traffic or at traffic lights. Computers monitor the engine and, if it is forced to idle when the vehicle stops, systems operate to temporarily shut down the engine, or the kinetic energy of the idling engine is used to power a small dynamo that charges a battery, which is later used to power the vehicle's electrical systems.

● Using new lightweight materials to replace heavier, traditional materials, such as steel, for vehicle parts. Reducing the mass of the car reduces its inertia to movement and so less energy is needed to get the vehicle moving in the first place.

Improving vehicle safety

One of the factors that affects the amount of damage to drivers and passengers in a car crash is the rapid deceleration that occurs when a car crashes. If you are in a car that suddenly slows down, your motion keeps you moving forward until a force acts to change your velocity (Newton's first law) – this could be the force between your head and the windscreen!

From Newton's second law, where:

force (N) = mass (kg) × acceleration (m/s²)

if you decelerate very quickly, there will be a large force on your body. Anything that increases the time taken for the collision, and in so doing reduces the deceleration, reduces the force acting on you. The trick is to engineer systems within the car that increase the time for a collision and yet keep the passengers safe inside a strong passenger compartment. Car makers design their cars so that they will collapse gradually on impact (crumple zones) – significantly increasing the time of the collision (reducing the acceleration) and substantially reducing the force on the occupants.

Now test yourself

TESTED

8 List three ways to improve the efficiency of a car.
9 Explain how a crumple zone reduces the impact force on a driver during a head-on collision.

Answers online

> **Exam tip**
>
> A common extended-answer exam question asks you to explain how car safety systems keep a driver and any passengers safe during a collision. Seat belts, crumple zones and air bags are good examples to choose, and they all work by extending the collision time and reducing the forces that act during a collision.

Summary

● When a force acts on a moving body, energy is transferred although the total amount of energy remains constant.
● The equation for work is:

work = force × distance moved in the direction of the force

$$W = Fd$$

● Work is a measure of the energy transfer, so that work = energy transfer (in the absence of thermal transfer).
● An object can possess energy because of its motion (kinetic energy) and position (gravitational potential energy) and deformation (elastic energy).
● The equation for kinetic energy is:

kinetic energy = A fixed resistor with a resistance of $0\frac{1}{2}$ × mass × (velocity)² of $25\,\Omega$ has a current of

$$KE = \frac{1}{2}mv^2$$

● The equation for change in potential energy is:

change in potential energy = mass × gravitational field strength × change in height

$$PE = mgh$$

● The relationship between force and extension for a spring is:

force = spring constant × extension

$$F = kx$$

● The work done in stretching can be determined by finding the area under the force–extension $(F–x)$ graph; $W = \frac{1}{2}Fx$ for a linear relationship between F and x.
● The energy efficiency of vehicles can be improved by reducing aerodynamic losses/air resistance and rolling resistance, idling losses and inertial losses.
● Seat belts, air bags and crumple zones on vehicles all work by extending the time of a collision and reducing the force acting on the occupants of the vehicle.

Exam practice

1 Figure 30.4 shows a low-loader lorry winching a car up a ramp. The winch of the lorry does 2450 J of work in lifting the car and 350 J of work against friction while pulling the car up the 3.5 m ramp.

Figure 30.4

a) Calculate the total work done in raising the car onto the back of the lorry. [1]

b) Select and use a suitable equation to find the force, F. [3]

(WJEC GCSE Physics P2 Foundation Tier Summer 2010 Q7)

2 Figure 30.5 shows a winch at Y which is used to pull a yacht at X, 50 m up a slipway, through a vertical height of 4 m.

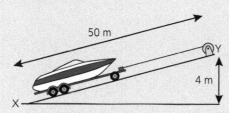

Figure 30.5

a) The weight of the yacht is 15 000 N, and it is lifted through the vertical height of 4 m. Select and use a suitable equation to calculate the work done against gravity, lifting the yacht through 4 m. [2]

b) A frictional force of 1000 N acts on the yacht as it is pulled up the 50 m slipway. Use your equation from (a) to calculate the work done against this frictional force. [1]

c) i) Hence, calculate the total amount of work done by the winch in pulling the yacht up the slipway. [1]

ii) Calculate the force that must be applied by the winch in pulling the yacht up the slipway. [2]

(WJEC GCSE Physics P2 Higher Tier Summer 2007 Q7)

3 A lift takes people up to a jump platform in a bungee tower. The jump platform is 55 m above the ground.

a) The lift takes a 60 kg person from the ground to the jump platform. Select a suitable equation and use it to find the increase in gravitational potential energy of the person. (Gravitational field strength = 10 N/kg) [3]

b) The bungee jumper has a kinetic energy of 18 000 J when he is falling at maximum speed.

i) What is his potential energy when he reaches his maximum speed? [1]

ii) Select and use an equation to find his maximum speed. [3]

c) Explain in terms of named forces why the speed increases before the bungee rope starts to stretch. [2]

d) The bungee rope stretches and stops the jumper just above ground level, storing the bungee jumper's energy in the rope. Give the values at this point of:

i) his kinetic energy [1]

ii) his gravitational potential energy [1]

iii) the energy stored in the bungee rope. [1]

(WJEC GCSE Physics P2 Higher Tier January 2008 Q5)

Answers and quick quiz 30 online

ONLINE

31 Stars and planets

The scale of the Universe

The Solar System

The Universe is a very big place – it would take about 13.75 thousand million (13.75 billion) years for light to travel from the Earth to the edge of the observable Universe. Our local patch of the Universe is called the Solar System. The main constituents of the Solar System are:

- 1 star – the Sun
- 8 planets – Mercury, Venus, Earth, Mars, Jupiter, Saturn, Uranus and Neptune
- 146 moons (a moon is a natural satellite of a planet)
- 5 dwarf planets, including Pluto
- an asteroid belt – between Mars and Jupiter
- many comets and other small lumps of rock and interplanetary dust.

Planets

Of the eight planets, the inner four are rocky planets (Mercury, Venus, Earth and Mars), and the outer four are gas giants (Jupiter, Saturn, Uranus and Neptune). Of the four rocky planets, Venus, Earth and Mars have atmospheres, and the gas giants are mostly composed of hydrogen, helium and some methane. Data on the eight planets is given in the table below.

Planet	Symbol	Mean orbit radius (in AU)	Orbital period (in Earth years)	Mean radius in $(R_\oplus)$	Mass in $(M_\oplus)$
Mercury	☿	0.39	0.24	0.38	0.06
Venus	♀	0.72	0.62	0.95	0.82
Earth	⊕	1.0	1.0	1.0	1.0
Mars	♂	1.5	1.9	0.53	0.11
Jupiter	♃	5.2	12	11	320
Saturn	♄	9.6	29	9.5	95
Uranus	♅	19	84	4.0	15
Neptune	♆	30	170	3.9	17

Table 31.1 Data on the planets of the Solar System.

Now test yourself

1 Explain how the four inner planets are different from the four outer planets.
2 What is the relationship between the orbital radius and orbital period for the planets?

Answers online

> **Exam tip**
>
> You need to learn the order of the planets away from the Sun. My Very Easy Method Just Speeds Up Naming!

Measuring distances in the Universe

- Earth radius, $R_\oplus$ – the size of a planet is measured relative to the Earth, so the radius of Jupiter = $11\,R_\oplus$. Comparisons to the Earth's dimensions are good measurements to use to compare the planets.
- Astronomical Units, AU – this is the average distance of the Earth from the Sun. Distances in the Solar System are measured using this unit. Neptune, the furthest planet, is 30 AU from the Sun and the outer reaches of the Solar System stretch out to over 100 000 AU. ($1\,\text{AU} = 1.5 \times 10^{11}\,\text{m}$)
- Light years, ly – the light-year is the distance that light travels in one year – $9.47 \times 10^{15}\,\text{m}$. This unit is used to measure distances to our nearest stars and within our own galaxy of stars, the Milky Way. The closest star to the Sun, Proxima Centauri, is 4.2 ly away. Our Solar System is about 4 ly in diameter, and the Milky Way galaxy is about 100 000 ly across. Our galaxy is part of a 'Local Group' of galaxies, about 10 million ly across, and the Local Group is part of the Virgo Supercluster of galaxy clusters, about 110 million ly across. The Virgo Supercluster is one of the largest observed structures in the Universe. The edge of the observable Universe is 13 750 million ly away.

> **Exam tip**
>
> You do not need to learn the conversion factors for metres into AU and ly. You would be given these values in the question paper.

Now test yourself

TESTED ▢

3 What is the distance of Jupiter away from the Sun:
   a) in AU
   b) in m
   c) in ly?

Answers online

How did the Solar System form?

The Sun and the Solar System formed out of the nebula (a gas and dust cloud) that resulted from the supernova death of a huge star. As our original nebula collapsed in on itself due to gravity, denser, darker regions appeared, where protostars formed. A protostar is part of a nebula, collapsing due to gravity, and it is the stage in a star's formation before nuclear fusion starts.

As the protostar collapsed further due to gravity, more gas and dust were drawn into it from the surrounding nebula. The pressure inside its core rose enough for the temperature to exceed 15 million °C and nuclear fusion reactions of hydrogen gas started and a star was born.

⊕Stellar lifecycles and the Hertzsprung–Russell diagram

REVISED ▢

The vast majority of stars spend most of their lives as **main-sequence stars**. The term 'main-sequence' was first coined by the Danish astronomer Ejnar Hertzsprung in 1907 who realised that the colour (or spectral class) of a star correlated to its apparent brightness (the brightness of a star as seen from Earth) and that many stars appeared to follow a simple relationship between these two variables. At the same time, an American astronomer called Henry Norris Russell was studying how the spectral class varied with actual (or absolute) brightness by correcting the brightness of stars for their distance away from Earth. The diagram showing absolute stellar brightness (or luminosity) against stellar temperature (which dictates the spectral class or colour of a star) is now known as the **Hertzsprung–Russell (HR) diagram** (Figure 31.1).

> A **main-sequence star** is one that releases energy by fusing hydrogen into helium.
>
> The **Hertzsprung–Russell (HR) diagram** is a means of displaying the properties of stars and depicting their evolutionary paths.

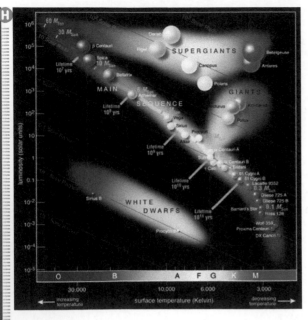

Figure 31.1 The Hertzsprung–Russell (HR) diagram.

The HR diagram arranges stars into groups according to their properties. The axes of the HR diagram in Figure 31.1 are luminosity (or the total amount of light energy emitted by the star, in solar units, where the luminosity of the Sun = 1) on the y-axis, and temperature (in kelvin) displayed as a non-linear power (of ten) series along the x-axis. The x-axis (which, unusually for a graph, runs backwards in temperature from left to right) can also be displayed as colour – red stars being coolest and blue stars being hottest. The HR diagram can be divided into four distinct quadrants, shown in Figure 31.2.

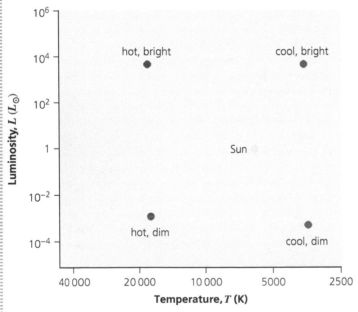

Figure 31.2 The four quadrants of the Hertzsprung–Russell (HR) diagram.

The hottest, brightest stars on the HR diagram are the largest main-sequence stars. The cool, brightest stars are red super-giants. The hot, dim stars are all white-dwarf stars, and the coolest, dim stars are red-dwarf stars.

Exam tip

Be careful – HR diagrams can have different axes, but look the same. The x-axis is normally temperature, but back to front, and as a power of 10 series. The y-axis can be luminosity or brightness (generally compared to the luminosity or brightness of the Sun) or absolute magnitude.

(H) **Now test yourself**

TESTED

4 What are the axes of a HR diagram?
5 Which groups of stars have the following properties:
  a) cool, dim stars
  b) hot, dim stars
  c) hot, bright stars
  d) cool, bright stars?
6 Where are main-sequence stars found on the HR diagram?

Answers online

Stellar lifecycles

Main-sequence stars run from top left to bottom right on the HR diagram. Above the main-sequence stars are the giant stars, which have a radius of between 10 and 100 times that of the Sun. A **red giant** is a dying star in one of the last stages of its evolution. A main-sequence star swells up to form a red giant when it starts to run out of hydrogen fuel and starts to fuse helium gas. Red giants are an important stage in the lifecycle of most main-sequence stars. Most stars spend most of their lifetime on the main sequence where their stability depends on the balance between the gravitational force of attraction trying to implode the star, and the combination of gas and radiation pressure trying to push the star outwards. Radiation pressure is the effect of the electromagnetic radiation moving out from the core of the star.

> A **red giant** is a very large star which fuses helium into heavier elements.
>
> A **white dwarf** is a star at the end of its life. No fusion occurs, it is cooling down.

When the hydrogen nuclear fuel of the main-sequence star starts to run out, nuclear fusion of helium takes over in the core. The radiation pressure increases and the balance of stability of the star is tipped in favour of the star expanding. As it gets bigger, the energy produced by the star is spread over a much larger surface area; its surface temperature drops; its colour becomes redder and it becomes a red giant. Red-giant stars are unstable, and the star increasingly relies on the nuclear fusion of heavier and heavier elements (a process called nucleosynthesis). Once the fusion reactions have produced the element iron, the star cannot gain energy from forming heavier elements and fusion ceases. The star collapses; its outer atmosphere is puffed outwards as a planetary nebula and the remaining hot core is termed a **white dwarf**. White-dwarf stars are found in the bottom left quadrant of the HR diagram. A white dwarf no longer undergoes nuclear fusion, but gives out light because it is still very hot.

The rest of the star's lifetime is a cooling process, as it is no longer generating energy via nuclear fusion. The white dwarf cools, moving to the right on the HR diagram, forming a red dwarf, and ultimately a black dwarf. The lifecycle plot of the Sun on the HR diagram, showing how its luminosity and surface temperature will change over its lifecycle, is shown in Figure 31.3.

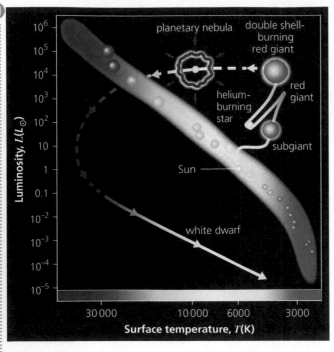

Figure 31.3 The HR diagram lifecycle plot of the Sun.

> **Exam tip**
>
> You need to learn the lifecycle of the Sun: protostar → main sequence → red giant → planetary nebula → white dwarf → red dwarf → black dwarf.

Supernovae, neutron stars and black holes

Massive main-sequence stars, with masses of up to 60 times that of the Sun, do not follow the main-sequence lifecycle path – they start their death process by swelling up to form a **supergiant** star. When nucleosynthesis stops, the supergiant undergoes rapid collapse and the resulting explosion is called a **supernova**. During the supernova collapse, the energy released is so great that elements heavier than iron are formed. All elements present in the Universe, heavier than iron, were once part of a supernova explosion. What is left after the supernova explosion depends on the final mass of the supergiant. 'Low'-mass supergiant remnants form huge nebulae, containing all the gas and dust required to start stellar formation once again. 'High'-mass supergiant remnants form **neutron stars**, where the material that made up the core of the supergiant is compressed into a space with a radius of about 12 km. Many neutron stars rotate at high speed forming pulsars which emit huge 'beams' of electromagnetic radiation, such as X-rays and gamma rays, as they do so.

The 'super-high' mass supergiant remnants are called **black holes**, formed from the cores of huge stars with a core mass about 10 times the mass of the Sun. These objects are compressed into a space with a radius of about 30 km and the gravitational attraction of a black hole is so large that not even light can escape – hence the name 'black hole'. Figure 31.4 summarises the death paths of stars.

A **supergiant** is the first stage in the death of a massive main-sequence star.

A **supernova** is a gigantic explosion caused by runaway fusion reactions.

A **neutron star** is a very small dense star made out of neutrons.

A **black hole** is the most concentrated form of matter from which not even light can escape.

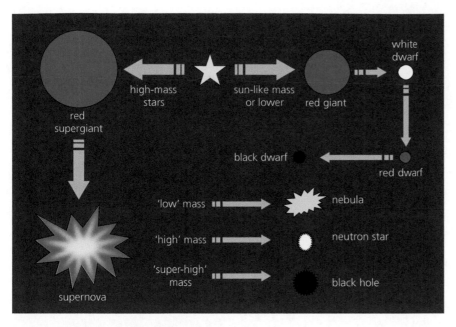

Figure 31.4 The death of stars.

Now test yourself

7 What is a supernova?
8 State the stages in the death of the Sun.
9 What are the similarities and differences between a neutron star and a black hole?

Answers online

Summary

- The Solar System consists of one star (the Sun), eight planets, several dwarf planets and many moons.
- Our local patch of space is called the Solar System, which is inside a galaxy called the Milky Way, which is part of a group of galaxies called the Local Group, which, in turn, is part of a cluster of 'groups' called the Virgo Supercluster.
- We need a range of distance scales when discussing the Universe: from the scale of planets and the Solar System, where comparisons to the Earth and the Sun are best; to the Milky Way galaxy and the observable Universe, where the distance that light travels in 1 year, called a light-year, is the best unit to use. $1\,ly = 9.47 \times 10^{15}\,m$.
- The astronomical unit, AU, is the mean distance of the Earth from the Sun. $1\,AU = 1.5 \times 10^{11}\,m$.
- Stars form from the gravitational collapse of nebulae. Protostars form out of high-density regions inside the nebula, before forming

main-sequence stars. Main-sequence stars, with a mass similar to that of the Sun, form red-giant stars before collapsing in on themselves, forming a planetary nebula and a white dwarf. Larger mass stars form super-giants before collapsing and exploding as a supernova, leaving a nebula, neutron star or black hole.
- The stability of stars depends on a balance between gravitational force and a combination of gas and radiation pressure; stars generate their energy by the fusion of increasingly heavier elements.
- Stellar material (including the heavy elements) is returned into space during the final stages in the lifecycle of giant stars.
- The Solar System was formed from the collapse of a cloud of gas and dust, including the elements ejected by a supernova.
- (H) The Hertzsprung–Russell (HR) diagram displays the properties of stars, and shows the evolutionary path of a star over the course of its lifetime.

Exam practice

1 The distances between objects in space are mind boggling. Even the distances between planets in our Solar System are so enormous that it takes space vehicles from Earth a very long time to get to them, many years in some cases. For example, the space craft called New Voyager that passed Pluto in 2015 was launched from Earth in January 2006 and, despite it being the fastest vehicle that has ever been sent from Earth, it took over 9 years to reach Pluto. Fortunately, there is one thing that travels so fast that we can express the vast distances of space in terms of how far it travels in 1 second or, for huge distances, the distance it travels in 1 year. This is of course light. Light travels 300000 kilometres in 1 second and even at that speed light takes 500s to travel to us from the Sun. We could say that the Sun is 500 light seconds away. The nearest star to our Sun is about 4 light years away, others are even millions of light years away from us.

Some of the distances used in astronomy are:
- 1 astronomical unit (AU) is the distance between the Earth and the Sun
- 1 light second is the distance travelled by light in 1 second = 300000km

a) i) What is a meant by a light-year? [1]

ii) The Sun is 500 light seconds away from Earth.
Use the equation below to calculate this distance in km. [2]

distance = speed × time

iii) The radius of Saturn's orbit is 9AU. Use your answer to part (ii) to calculate the radius of its orbit in km. [2]

b) When main-sequence stars come to the end of their 'lives' they go through stages which depend on their mass. Choose words or phrases from the list to complete the diagram that follows. [4]

red giant   black dwarf   white dwarf   supernova   neutron star

| red supergiant much more massive than our Sun |

↓

| |

↓

| |

| main-sequence star the size of our Sun |

↓

| |

↓

| |

(WJEC GCSE Physics Unit 2: Forces, space and radioactivity Foundation Tier SAM)

2 Our Solar System is made up of eight planets, each of which may or may not have a moon or moons orbiting them, many asteroids and some dwarf planets. The planets orbit the Sun. For planet Earth the orbit time is one year. The number of years that a planet takes to orbit the Sun depends on its distance from the Sun in the way shown in the table below. The table gives data on six of the planets in the Solar System.

Planet	Mean distance from the Sun (× 10^8 km)	Mean surface temperature (°C)	Time for one orbit of the Sun (years)
Venus	1.1.0	480	0.62
Earth	1.50	22	1.00
Mars	2.25	−23	1.88
Jupiter	7.80	−150	11.86
Saturn	14.00	−180	29.46
Uranus	29.00	−210	84.01

The graph in Figure 31.5 shows how the orbital speed of the planets changes with their distance from the Sun.

→

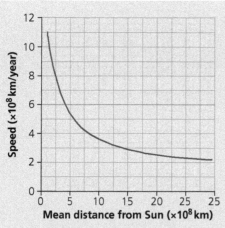

Figure 31.5

Use data from the table and graph to answer the following questions.

a) What is the orbital speed of Saturn (in km/year)? [1]

b) A dwarf planet, Ceres, is 700 km in diameter and has an orbital speed of 5.8×10^8 km/year. It travels 2.67×10^9 km in making one orbit of the Sun.

   i) Use the graph to find the distance of Ceres from the Sun. [1]

   ii) Use the equation below to calculate the orbital time of Ceres. [2]

$$\text{time} = \frac{\text{distance}}{\text{speed}}$$

c) Estimate the mean temperature on Ceres, show your working or explain how you arrived at your answer. [2]

d) State two reasons why Ceres takes longer than Earth to complete one orbit of the Sun. [2]

(WJEC GCSE Physics Unit 2: Forces, space and radioactivity Higher Tier SAM Q4)

3 The Solar System consists of the Sun and its planets.

a) Name the force that keeps the planets in orbit around the Sun. [1]

b) i) Apart from the Earth, name one planet that has a rocky structure. [1]

   ii) Name two planets that have a gas structure. [1]

c) The table below gives data on four planets in the Solar System.

The asteroid belt lies between Mars and Jupiter. Asteroids are bits of rock, of varying size, which never collected to form a planet.

If a planet had formed from the bits of rock, use the data in the table to estimate its:

   i) distance from the Sun

   ii) orbit time

   iii) surface temperature. [3]

Planet	Mean distance from the Sun ($\times 10^8$ km)	Mean surface temperature (°C)	Time for one orbit of the Sun (years)
Earth	1.50	22	1.00
Mars	2.25	–23	1.88
Jupiter	7.80	–150	11.86
Saturn	14.00	–180	29.46

(WJEC Physics P1 Higher January 2007 Q3)

4 Our Sun was created and will eventually die over billions of years. The sentences below describe the stages in its life.

A The Sun goes through a stable state.

B The Sun shrinks to become a white dwarf.

C Gravity pulls dust and gas together.

D The Sun becomes a red giant.

   a) Put the letters A, B, C and D in the correct order. [3]

   b) In which stage, A, B, C or D, is the Sun at present? [1]

(WJEC GCSE Physics P3 Foundation Tier Summer 2010 Q1)

Answers and quick quiz 31 online

ONLINE

32 Types of radiation

Inside the nucleus

The nucleus of an atom contains positively charged particles, protons, and neutral particles, neutrons. The number of protons in the nucleus is called the **proton number**, Z; the number of protons plus the number of neutrons is called the **nucleon number**, A. The values of Z and A are often shown using the $_Z^A X$ notation, where X is the chemical symbol for the atom in question. For example, 52.4 per cent of all naturally occurring lead atoms have nuclei made up of 82 protons and 126 neutrons, a total of 208 nucleons, i.e. $_{82}^{208}Pb$. Lead also has other **isotopes** – nuclei with the same number of protons, but different numbers of neutrons. The different isotopes are often written as Pb-208, Pb-207, etc., where the number refers to the nucleon number.

> **Proton number** is the number of protons.
>
> **Nucleon number** is the number of protons and neutrons.
>
> **Isotopes** are different forms of a particular element. Isotopes have the same number of protons but different numbers of neutrons.

Now test yourself

1 Use the $_Z^A X$ notation to describe the following radioactive nuclei in the table.

Nucleus	Proton number, Z	Number of neutrons	$_Z^A X$
Lithium	3	4	
Carbon	6	7	
Strontium	38	52	
Technetium	43	56	

Answers online

> **Exam tip**
>
> The proton number is sometimes called the atomic number, and the nucleon number is sometimes called the atomic mass. These values are not exact, because they describe atoms not nuclei.

> **Exam tip**
>
> You can easily work out the number of neutrons in a nucleus by subtracting the proton number from the nucleon number.

Nuclear radiation

Some types of atom are radioactive. This means that the nucleus of the atom is unstable, due to an imbalance of protons and neutrons, and it can break apart, emitting ionising radiation in the form of alpha (α), beta (β) or gamma (γ) radiation.
- Alpha particles are helium nuclei. They are the most ionising (and, therefore, cause the most harm inside the body to living cells) and the least penetrating type of nuclear radiation – they are absorbed by a thin sheet of paper or by skin. Alpha-emitting nuclear waste is easily stored in plastic or metal canisters.
- Beta particles are high-energy electrons. They have medium ionising ability (and cause little harm inside the body to living cells); they

are absorbed by a few millimetres of aluminium or Perspex plastic, and beta-emitting nuclear waste is stored inside metal canisters and concrete silos.

- Gamma rays are high-energy electromagnetic waves. They are the least ionising (about 20 times lower than alpha particles) and, therefore, cause the least harm to living cells inside the body. They are the most penetrating, able to travel through several centimetres of lead. Nuclear waste containing gamma emitters needs storage inside lead-lined, thick concrete silos.

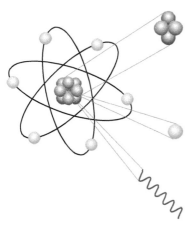

Alpha (α) radiation
These are particles, not rays, and travel at about 10% of the speed of light. An α particle is identical to a helium nucleus, consisting of 2 protons and 2 neutrons joined together.

Beta (β) radiation
These are fast-moving electrons that come from the nucleus. They travel at about 50% of the speed of light.

Gamma (γ) radiation
This is an electromagnetic wave. It travels at the speed of light (3×10^8 m/s). It has very high energy.

Figure 32.1 Alpha, beta and gamma radiation.

Now test yourself

TESTED

2 Complete the summary table of the three different types of ionising radiation.

Type of radiation	Symbol	$^A_Z X$	Penetrating power	Ionising power
Alpha				
Beta				
Gamma				

Answers online

Stability of the nucleus

REVISED

In a stable atom there is an optimum balance between the number of protons and the number of neutrons in the nucleus. But some isotopes can have too few or too many neutrons, causing an imbalance and making the nucleus unstable and radioactive. For example:

- the lead isotope Pb-181 has only 99 neutrons and is an alpha particle emitter
- the lead isotope Pb-214 has 132 neutrons and is a beta emitter. The nucleus becomes more stable by emitting alpha or beta particles, restoring the optimum balance of protons and neutrons, and sometimes emitting gamma radiation too.

Nuclear equations

The $^A_Z X$ notation can be used to represent the decay of radioactive nuclei in a nuclear equation. Alpha particles are written as $^4_2 He$ because they consist of two protons and two neutrons, like a helium nucleus. Beta particles are written as $^0_{-1} e$, because they are electrons.

Alpha decay

The nuclear decay equation for the alpha decay of lead-181 is:

$$^{181}_{82} Pb \rightarrow {}^4_2 He + {}^{177}_{80} Hg$$

The lead-181 nucleus emits an alpha particle, losing 4 nucleons (2 protons + 2 neutrons), forming mercury-177.

Beta decay

The nuclear decay equation for the beta decay of lead-214 is:

$$^{214}_{82} Pb \rightarrow {}^0_{-1} e + {}^{214}_{83} Bi$$

The lead-214 nucleus emits a beta particle (electron). The nucleon number stays the same, but the proton number goes up by one, forming bismuth-214.

> **Exam tip**
>
> All nuclear equations must balance: the total proton number on each side must be the same, and the total nucleon number on each side must be the same.

Now test yourself

3 Complete the following nuclear decay equations, by determining A and Z:

a) $^{241}_{95} Am \rightarrow {}^4_2 He + {}^A_Z Np$

b) $^{225}_{88} Ra \rightarrow {}^0_{-1} e + {}^A_Z Ac$

Answers online

Background radiation

Background radiation is all around us and it comes naturally from our environment and from artificial (human-made) sources. The background-radiation count rate needs to be subtracted from any measurements made of nuclear radiation. The pie chart in Figure 32.2 shows the mean contribution of the different sources to our background radiation.

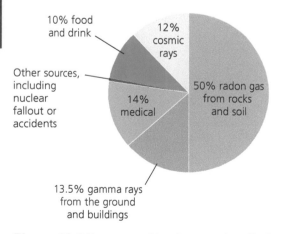

Figure 32.2 Sources of background radiation.

Most of the background radiation comes from naturally occurring sources, primarily from the ground, rocks and from space. Artificial background radiation mostly comes from medical sources, predominantly as a result of medical and dental examinations using X-rays. The largest source of the background radiation comes from the radioactive element radon, emitted from rocks (granite) and the soil. Radon is a gas, and it can escape from granite and is easily breathed in by humans, entering our lungs, where it can decay. The alpha particles emitted by decaying radon are absorbed by the cells lining the lungs, causing the cells to die or mutate (forming cancers).

Monitoring and measuring radioactive decay

REVISED

Radioactive decay is a random process and this has consequences when undertaking experimental work. Readings should be repeated and mean averages taken; background radiation must be subtracted from the readings, and the readings generally need to be taken over a lengthy period. All of these techniques are used to reduce the effect of random fluctuations in the measurements.

Now test yourself

TESTED

4 What is background radiation?
5 How are radioactivity measurements improved by taking into account the fact that they are random processes?

Answers online

Nuclear waste

REVISED

High-level nuclear waste from nuclear reactor cores is stored temporarily in water, surrounded by lots of concrete and lead shielding, while it cools. The radiation is absorbed by the water, concrete and lead. In the longer term, this waste is encased in blocks of glass – a process known as 'vitrification'. The glass blocks are then stored deep underground, where the surrounding rocks absorb the radiation. The initial cooling and vitrification processes take decades and the radioactive material stays radioactive for millions of years. Intermediate nuclear waste from medical processes is less radioactive than high-level waste and is mixed with concrete and poured into steel drums, before being stored securely underground.

> **Exam tip**
>
> Question 6 is an example of a question containing complex tables. In the exam you must spend time studying tables very carefully. Re-read the column and row headings and make sure that you know exactly what data is being given to you.

Now test yourself

TESTED

6 Some radioactive elements emit more than one type of radiation.
The apparatus in Figure 32.3 was used to investigate the radiation emitted from three sources, A, B and C. The sources were always placed at the same position, close to the detector. The table below shows the mean counts per minute obtained when different materials were placed between the sources and the detector. All the readings have been corrected for background radiation.

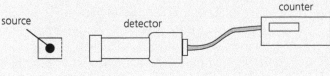

Figure 32.3

	Mean counts/min with nothing between source and detector	Mean counts/min with thin paper in the way	Mean counts/min with 3 mm of aluminium in the way	Mean counts/min with 2 cm of lead in the way
A	256	256	256	85
B	135	80	80	0
C	310	310	188	0

a) How can you tell that source A is emitting gamma (γ) radiation?

b) Which source, A, B or C, emits alpha (α) particles? Give a reason for your answer.

c) The beta radiation source contains atoms of strontium-90, $^{90}_{38}$Sr.

   i) Explain what happens to a Sr-90 atom when it decays.

   ii) The elements nearest to strontium in the periodic table are shown below. Use the information in the table to determine the daughter nucleus produced when Sr-90 decays by beta emission.

Element	Krypton	Rubidium	Strontium	Yttrium	Zirconium
Symbol	Kr	Rb	Sr	Y	Zr
Proton number	36	37	38	39	40

Answers online

Summary

- Substances that are radioactive can emit alpha (α; $^4_2$He), beta (β; $^{\ 0}_{-1}$e) and gamma (γ) radiation.
- Alpha (α), beta (β) and gamma (γ) radiation, ultraviolet light and X-rays are all types of ionising radiation. Ionising radiation is able to interact with atoms and damage cells because of the energy it carries.
- Radioactive emissions from unstable atomic nuclei arise because of an imbalance between the numbers of protons and neutrons.
- The number of protons and neutrons in an atomic nucleus is called the nucleon number or the mass number, (A), and the number of protons is called the proton number, (Z); chemists usually call it the atomic number.
- The nuclear symbols in the form of the ^A_ZX notation (where X is the atomic symbol from the periodic table) are used to describe radioactive atoms, decays and balanced radioactive decay equations.
- The waste materials from nuclear power stations and nuclear medicine are radioactive; some of them will remain radioactive for thousands of years.
- When measurements of radiation are taken, an allowance for background radiation must be made.
- Alpha, beta and gamma radiation have different penetrating powers. Alpha radiation is absorbed by a thin sheet of paper, beta radiation is absorbed by a few millimetres of aluminium or Perspex, but gamma radiation is only absorbed by a few centimetres of lead.
- The differences in the penetrating power of alpha, beta and gamma radiation determine

their potential for harm. Alpha radiation is easily absorbed but is the most ionising. Gamma rays are very penetrating but are about 20 times less ionising than alpha radiation.

- Radioactive waste is stored in a series of containment systems. Steel canisters, water, concrete, glass and lead are all used to shield the environment from harmful doses of radiation. The radiation produced by the waste is absorbed by the different types and thicknesses of containment.
- The long-term solution to the storage of radioactive waste is deep underground, where the harmful radiation can be absorbed by the surrounding rocks.
- Background radiation is all around us and comes from natural or artificial (human-made) sources.
- Natural sources of background radiation include radon from rocks, gamma rays from the ground and buildings, cosmic rays from space and radiation contained in food and drink.
- Artificial sources of background radiation include X-rays from medical examinations and nuclear fallout from weapons tests or accidents.
- Most of our background radiation (between 50 per cent and 90 per cent) comes from radon gas (depending on where you live). Places like Cornwall, where there is a lot of granite rock, have higher levels of radon gas because the granite contains uranium that decays (eventually) to radon.

Exam practice

1 Read the information below.
The first 92 elements in the periodic table occur naturally on the Earth. Other elements have been created by mankind, usually inside nuclear reactors. The atoms of some elements exist in different forms, which are called isotopes. Isotopes of the same element all have the same number of protons. However, isotopes of different elements may have the same nucleon number, some of which are shown in the table below.

Isotope	Proton number	Nucleon number
Americium Am)	95	238
Uranium (U)	92	238
Thorium (Th)	90	238
Californium (Cf)	98	238

a) Read the statements below and tick (✓) the correct ones. [3]

Statement	
Atoms of all of these isotopes have the same number of protons in their nuclei.	
An atom of uranium has 92 neutrons in its nucleus.	
An atom of californium has the greatest number of protons in its nucleus.	
An atom of californium has the smallest number of neutrons in its nucleus.	
Uranium is not a naturally occurring element.	
An atom of uranium has 92 protons in its nucleus.	

b) Complete the decay equation of uranium-238 into thorium in the equation below. [2]

$$^{238}_{92}U \rightarrow {}^{4}_{2}\alpha + {}^{-}_{-}Th$$

c) Identify the two correct isotopes of uranium from the list below. [2]

$^{238}_{92}U \qquad ^{238}_{89}U \qquad ^{234}_{90}U \qquad ^{235}_{92}U \qquad ^{238}_{91}U$

(WJEC GCSE Physics Unit 2: Forces, space and radioactivity Foundation Tier SAM Q1)

2 The sources and percentages of background radiation are shown in the pie chart in Figure 33.4. The pie chart is not drawn to scale.

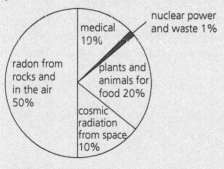

Figure 32.4

a) i) What percentage of background radiation comes from natural sources? [1]
   ii) Use the pie chart to explain why an increase in nuclear-power generation would only produce a small increase in the background radiation. [1]
b) Give a reason why background radiation varies from place to place. [1]

(WJEC GCSE Physics P2 Foundation Tier Summer 2010 Q4)

➜

3 A piece of rock gives a reading on a counter which is connected to a radiation detector. The rock is wrapped in various materials and different readings in counts per minute (cpm) are observed.

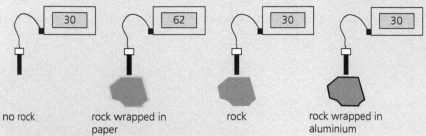

| 30 | 62 | 30 | 30 |

no rock   rock wrapped in   rock   rock wrapped in
                 paper   aluminium

Figure 32.5

a) i) State the type, or types of radiation that the rock emits.   [2]
   ii) Explain your choice.   [2]
b) The rock is now wrapped in lead. Explain what you would expect the reading on the counter to be.   [2]

(WJEC/EDUQAS GCSE Physics Component 1 Concepts in Physics Foundation Tier Sample Paper Q8)

4 a) i) Give a reason why radioactive waste is a health risk to the population, e.g. a cancer risk.
     ii) Give a reason why radioactive waste is expensive to dispose of safely.   [2]
   b) A sample of radioactive waste emits alpha (α), beta (β) and gamma (γ) radiation. When placed in front of a detector the sample gives a reading of 450 counts/minute. The diagrams in Figure 32.6 show the count rate when different absorbers are placed between the sample and the detector.

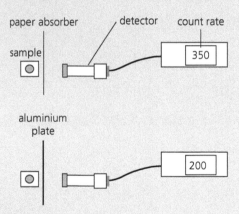

paper absorber   detector   count rate

sample

| 350 |

aluminium
plate

| 200 |

Figure 32.6

i)   How much of the 450 counts/minute is due to alpha (α) radiation?   [1]
ii) How much of the 450 counts/minute is due to gamma (γ) radiation?   [1]
iii) Explain why the aluminium absorber reduces the original count rate by 250 counts/minute. [2]

(WJEC GCSE Physics P2 Foundation Tier Summer 2010 Q8)

Answers and quick quiz 32 online

ONLINE

33 Half-life

Radioactive decay and probability

Radioactive decay is a random process, but all radioactive substances decay in a similar way, following the same pattern. Radioactive atoms have a constant probability of decay, and follow the same rules of probability that other random processes follow, such as throwing dice. The probability of throwing a six on a dice is 1/6, in the same way that a radioactive atom has a constant numerical probability of decay. This means that, although you cannot say for definite which atoms of a radioactive substance will decay in a set time, you can predict how many of them will decay. Each radioactive substance has a given probability of decay; some have very high probabilities and decay very quickly indeed, whereas others have extremely small probabilities and remain radioactive for very long times. The unit of radioactive decay activity is the becquerel, Bq. An activity of 1 Bq is equivalent to 1 radioactive decay per second.

Half-life

Radioactive materials decay in a very predictable way. The graph in Figure 33.1 shows a typical radioactive decay curve, for the decay of iridium-192, Ir-192.

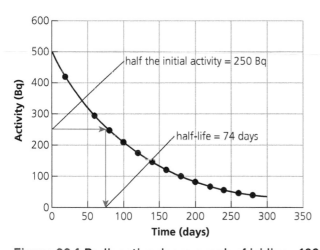

Figure 33.1 Radioactive decay graph of iridium-192.

The initial activity of the sample is 500 Bq. After 74 days the activity has fallen to 250 Bq – half the original amount. This time is called the half-life of iridium-192. After another 74 days, the activity has fallen to 125 Bq – half of 250 Bq – and after each half-life the activity halves. The half-life of a radioactive substance is the time taken for the activity of the substance to halve. Some substances have very short half-lives and decay extremely quickly, whereas others have very long half-lives and remain radioactive for very long times – some have half-lives considerably longer than the age of the Universe.

An iridium-192 source has a half-life of 74 days and an initial activity of 1200 Bq. What will the activity of the iridium source be after 222 days?

Answer

Number of half-lives in 222 days = $\dfrac{222 \text{ days}}{74 \text{ days}}$ = 3 half-lives

After 1 half-life (74 days) the activity will be $\dfrac{1200 \text{ Bq}}{2}$ = 600 Bq

After the second half-life (148 days), the activity will be $\dfrac{600 \text{ Bq}}{2}$ = 300 Bq

After the third half-life (222 days), the activity will be $\dfrac{300 \text{ Bq}}{2}$ = 150 Bq

Exam tip

The decay of every radioactive material follows the same pattern. The activities and the half-lives may be very different, but the shape of the decay curve is always the same.

Now test yourself

TESTED

1 Why do all radioactive substances decay with a similar pattern?
2 What is meant by radioactive half-life?
3 To study blood flow, a doctor injects some technetium-99 (Tc-99) into a patient. The gamma radiation given out by the Tc-99 atoms is detected using a gamma camera outside the patient's body. The graph in Figure 33.2 shows how the count rate from a sample of Tc-99 changes with time.

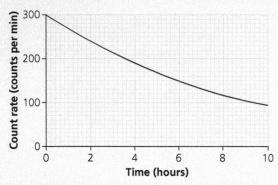

Figure 33.2

a) i) How many hours does it take for the count rate to fall from 300 counts per minute to 150 counts per minute?
   ii) What is the half-life of Tc-99?
   iii) How long will it take for the count rate to fall from 300 to 75 counts per minute?
b) Explain why an alpha-emitting source would be unsuitable to study blood flow.

Answers online

Using radioactivity

REVISED

The uses of radioactive materials depend on their properties, in particular:
● half-life
● penetrating power
● ionising power.

In medicine, radioactive materials are used in two main ways: in imaging and in therapy (treatment). The isotope carbon-14 can also be used to date the age of very old (dead) materials.

Radio-imaging

The radioactive material is injected into the body. It makes its way to the particular place that requires investigation and it emits gamma rays, which can be detected outside of the body. Gamma emitters are used, such as technetium-99 with a half-life of six hours. This will only remain radioactive for 30 hours or so (about 5 half-lives). Gamma rays cause few problems to the body as they pass straight through and are very weak ionisers.

Radiotherapy

This involves the use of radioactive materials to kill affected (usually cancerous) cells. Beta radiation may be used, as this has a short range in flesh, so that it only damages the cells close to the target area. The half-life chosen is typically a few days, but depends on the dose required – longer half-lives will deliver higher doses.

Carbon dating

Carbon-14 is a naturally occurring radioactive isotope of carbon. Only about 1 atom in every 10 000 000 000 carbon atoms is an atom of carbon-14. Carbon-14 is a radioactive beta emitter with a half-life of 5730 years and it can be used to date organic objects up to about 60 000 years old – about the time when our early ancestors, *Homo sapiens*, started to migrate out of Africa. All living things contain carbon, and the ratio of non-radioactive carbon-12 atoms to radioactive carbon-14 atoms in living material is known very precisely – as it depends on the composition of the carbon dioxide in the atmosphere. When an organic living creature or plant dies, the ratio of carbon-12 to carbon-14 starts to change as the carbon-14 decays and no more fresh carbon-14 is added (because photosynthesis and/or respiration is no longer carried out by the creature or plant once it is dead). If the ratio of carbon-12 to carbon-14 in a dead organic object is measured, then it is possible to use the half-life of carbon-14 to work backwards to find out when the ratio was the same as it is in living organisms now. A graph similar to the one in Figure 33.3 can be used to measure the percentage of carbon-14 remaining, compared with a living sample.

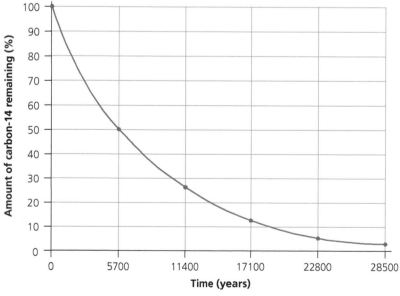

Figure 33.3 Graph of radioactive decay of carbon-14.

Exam tip

Question 6 below involves drawing a decay graph. When you have to draw a graph in an exam, you are nearly always given the axes and grid. The most important thing is to plot the points accurately and correctly, and draw a best-fit curve – always check your graph two or three times after you have plotted it and before you draw the curve, and use pencil so that you can rub it out if you make a mistake.

Now test yourself

4 The use of a radioactive substance depends on which three properties of the radioactive substance?
5 List two medical uses of radioactive substances.
6 a) Carbon-14 has a half-life of 5700 years. Sketch a graph of activity against time, showing the decay of carbon-14 from an initial activity of 64 counts per minute.
  b) While trees are alive they absorb and emit carbon-14 (in the form of carbon dioxide) so that the amount of carbon-14 in them remains constant.
    i) What happens to the amount of carbon-14 in a tree after it dies?
    ii) A sample of wood from an ancient dwelling gives 36 counts per minute. A similar sample of living wood has 64 counts per minute. From your graph, deduce the age of the dwelling. (Show on your graph how you obtained your answer.)

Answers online

Summary

- Radioactive decay is a random event, governed by the laws of probability. Radioactive decay can be modelled by rolling a large collection of dice, tossing a large number of coins or using a suitably programmed spreadsheet.
- The half-life is the length of time it takes for half the atoms in the sample to decay; this is a constant for a particular element. Half-lives vary between different radioactive elements from less than a second to billions of years.
- The activity of a sample of isotope can be plotted against time on a graph, and from this the half-life of that isotope can be measured. The graph is called a radioactive decay graph.

- The unit of radioactive decay is the becquerel, Bq. 1 Bq is 1 radioactive decay per second.
- Carbon-14 is a naturally occurring radioactive isotope of carbon. It emits beta particles with a half-life of 5370 years. By measuring the proportion of carbon-14 to the usual isotope carbon-12, organic objects up to around 60 000 years old can be dated.
- The characteristics of the different forms of radioactive decay, such as half-life, penetrating power and ionising ability, mean that they are useful for different purposes, for example medical uses such as radio-imaging and radiotherapy.

Exam practice

1 A student does an experiment with dice to investigate radioactive decay. The dice, which represent radioactive atoms, are thrown together onto the floor. Those that show a six are removed. These represent the atoms whose nuclei have decayed. The remaining dice (undecayed atoms) are thrown again and the process is repeated several times. The student starts with 600 dice.
  a) i) Predict how many of the dice would show a six on the first throw. [1]
    ii) State why the student cannot predict which dice will show a six. [1]
  b) The results of the experiment are shown in the table below.

Throw	Number of sixes	Number of dice remaining
0	0	600
1	95	505
2	85	420
3		350
4	60	290
5	50	240
6	40	200
7	30	170
8	25	145

    i) Fill in the gap in the table above. [1]
    ii) Plot the results on the grid below and draw a suitable line. [3]
      Three points have been plotted for you.

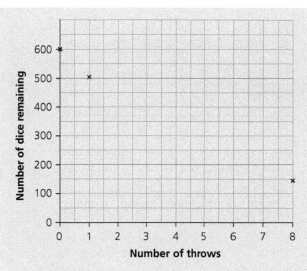

Figure 33.4

    iii) Draw lines on your graph to enable you to find the half-life of the dice. [2]

c) Americium-241 is a radioactive substance which is used in smoke alarms in houses. It decays by emitting alpha particles.

    i) State why Americium-241 is radioactive. [1]

    ii) What is an alpha particle? [1]

    iii) Explain why the use of Americium-241 in house smoke alarms, when in normal use, does not present a significant health risk to people living in the houses. [2]

(WJEC GCSE Additional Science/Physics P2 Higher Tier January 2016 Q1)

2 The graph in Figure 33.5 shows the radioactive decay in counts per minute (cpm) of a sample of carbon-14.

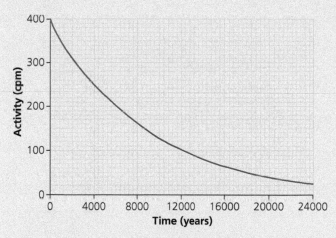

Figure 33.5

a) Use information from the graph to answer the following questions.

    i) State the activity after 4000 years. [1]

    ii) State the time taken for the activity to fall from 400 cpm to 100 cpm. [1]

    iii) State the half-life of carbon-14. [1]

    iv) State the time it would have taken for the activity to have fallen from 800 cpm to 400 cpm. [1]

b) The nuclear symbol for carbon-14 is C. Complete the following table for the nucleus of carbon-14. [3]

Nucleon number	
Number of protons in its nucleus	
Number of neutrons in its nucleus	

(WJEC GCSE Additional Science/Physics P2 Higher Tier May 2016 Q4)

→

3 Nuclear medicine uses radioisotopes which emit radiation from within the body. One tracer uses iodine, which is injected into the body to treat the thyroid gland. The table shows four isotopes of iodine.

Form of iodine	Radiation emitted	Half-life
Iodine-125	Gamma	59.4 days
Iodine-128	Beta	25 minutes
Iodine-129	Beta and gamma	15 000 000 years
Iodine-131	Beta and gamma	8.4 days

a) Iodine-129 emits both beta and gamma radiation. Describe the nature of these types of radiation. [2]
b) The table shows that the half-life of iodine-125 is 59.4 days. State what this means. [2]
c) i) Use the data to explain why iodine-131 is the most suitable form of iodine for treating thyroid cancer. [2]
   ii) Patients are advised that after treatment with iodine-131, the radiation they are exposed to will not drop to the background value until 12 weeks after treatment. Calculate the fraction of radioactivity due to iodine-131 remaining after 12 weeks. [3]

(WJEC GCSE Additional Science/PhysicsP2 Higher Tier May 2016 Q3)

4 Isotopes of iodine can be used to study the thyroid gland in the body. A small amount of the radioactive isotope is injected into a patient and the radiation is detected outside the body. Three isotopes that could be used are: $^{123}_{53}I$; $^{131}_{53}I$ and $^{132}_{53}I$. They have half-lives of 13.22 hours, 8 days and 13.2 hours respectively.
Answer the following question in terms of the numbers of particles.
a) Compare the structures of the nuclei of $^{123}_{53}I$ and $^{131}_{53}I$. [2]
b) The nucleus of $^{131}_{53}I$ decays into xenon (Xe) by giving out beta (β) and gamma (γ) radiation.
   i) What is beta radiation? [1]
   ii) Complete the equation below to show the decay of I-131. [2]

$$^{131}_{53}I \rightarrow \,^{...}_{54}Xe + \,^{0}_{...}\beta + \gamma$$

c) The isotope $^{123}_{53}I$ decays by gamma emission. Explain why it is better to use $^{123}_{53}I$ than $^{131}_{53}I$ as a medical tracer. [2]
d) i) I-131 has a half-life of 8 days. Explain what this statement means. [2]
   ii) Following the nuclear power-station disaster in Japan 2011, people living in the area were given non-radioactive iodine-127 ($^{127}_{53}I$) supplement tablets to reduce their intake of iodine-131 that leaked from the reactor. Calculate the length of time that people had to take the supplement before the activity of iodine-131 reduced to approximately 3 per cent of its original value immediately after the leak. [2]

(WJEC GCSE Physics Unit 2: Forces, space and radioactivity Higher Tier SAM Q7)

Answers and quick quiz 33 online

ONLINE

Equations

current = $\dfrac{\text{voltage}}{\text{resistance}}$	$I = \dfrac{V}{R}$
total resistance in a series circuit	$R = R_1 + R_2$
total resistance in a parallel circuit	$\dfrac{1}{R} = \dfrac{1}{R_1} = \dfrac{1}{R_2}$
energy transferred = power × time	$E = Pt$
power = voltage × current	$P = VI$
power = current² × resistance	$P = I^2R$
% efficiency = $\dfrac{\text{energy [or power] usefully transferred}}{\text{total energy [or power] supplied}} \times 100$	
density = $\dfrac{\text{mass}}{\text{volume}}$	$\rho = \dfrac{m}{V}$
units used (kWh) = power(kW) × time (h) cost = units used × cost per unit	
wave speed = wavelength × frequency	$v = \lambda f$
speed = $\dfrac{\text{distance}}{\text{time}}$	
pressure = $\dfrac{\text{force}}{\text{area}}$	$p = \dfrac{F}{A}$
p = pressure, V = volume, T = kelvin temperature	$\dfrac{pV}{T}$ = constant
	$T/K = \theta/°C + 273$
change in thermal energy = mass × specific heat capacity × change in temperature	$\Delta Q = mc\Delta\theta$
thermal energy for a change of state = mass × specific latent heat	$Q = mL$
force on a conductor (at right angles to a magnetic field) carrying a current = magnetic field strength × current × length	$F = BIl$
V_1 = voltage across the primary coil, V_2 = voltage across the secondary coil, N_1 = number of turns on the primary coil, N_2 = number of turns on the secondary coil	$\dfrac{V_1}{V_2} = \dfrac{N_1}{N_2}$
acceleration [or deceleration] = $\dfrac{\text{change in velocity}}{\text{time}}$	$a = \dfrac{\Delta v}{t}$
acceleration = gradient of a velocity–time graph	
distance travelled = area under a velocity–time graph	
resultant force = mass × acceleration	$F = ma$
weight = mass × gravitational field strength	$W = mg$
work = force × distance	$W = Fd$
kinetic energy = $\dfrac{\text{mass} \times \text{velocity}^2}{2}$	$KE = \dfrac{1}{2}mv^2$
change in potential energy = mass × gravitational field strength × change in height	$PE = mg$
force = spring constant × extension	$F = kx$
work done in stretching = area under a force–extension graph	$W = \dfrac{1}{2}Fx$
momentum = mass × velocity	$p = mv$
force = $\dfrac{\text{change in momentum}}{\text{time}}$	$F = \dfrac{\Delta p}{t}$
u = initial velocity v = final velocity t = time a = acceleration x = displacement	$v = u + at$ $x = \dfrac{u+v}{2}t$ $x = ut + \dfrac{1}{2}at^2$ $v^2 = u^2 + 2ax$
moment = force × distance	$M = Fd$